Multimedia Technology Basics

李 婷 吕建红 王寅龙 主编
刘笑飞 武 洁 刘 玲 等译

华中科技大学出版社
http://press.hust.edu.cn
中国·武汉

Brief Introduction of the Content

Multimedia technology is a very practical computer application technology. Mastering this technology can bring us a lot of convenience in our daily work and life. This book introduces the multimedia technology from three aspects: what multimedia technology is, how to process the commonly used multimedia materials such as image, audio, and video, and how to use multimedia materials to do multimedia works.

The first chapter introduces the basic knowledge of multimedia technology, including related concepts and commonly used multimedia materials; Chapter Two introduces the relevant knowledge of digital image processing, including the imaging principle of digital image, digital image correction, beautification, matting, and digital image synthesis; Chapter Three introduces the related knowledge of digital audio processing, including audio basic knowledge, audio recording, audio editing, and audio effect processing; Chapter Four introduces the related knowledge of digital video processing, including the basic knowledge of video, the editing of video production materials, the addition of video effects and subtitles, and the output of video; Chapter Five takes PowerPoint software as an example to introduce the related knowledge of multimedia platform software, including the types and functions of multimedia platform software, the basic application of multimedia materials in PowerPoint software, and the basic skills of PowerPoint design.

图书在版编目(CIP)数据

多媒体技术基础＝Multimedia Technology Basics：英文/李婷，吕建红，王寅龙主编；刘笑飞等译. —武汉：华中科技大学出版社，2023.5
ISBN 978-7-5680-8898-5

Ⅰ.①多… Ⅱ.①李… ②吕… ③王… ④刘… Ⅲ.①多媒体技术-英文 Ⅳ.①TP37

中国国家版本馆 CIP 数据核字(2023)第 058036 号

Multimedia Technology Basics	李　婷　吕建红　王寅龙　主编
	刘笑飞　等译

策划编辑：张　玲	责任校对：陈元玉
责任编辑：徐定翔	责任监印：周治超
封面设计：杨小勤	
出版发行：华中科技大学出版社(中国•武汉)	电　话：(027)81321913
武汉市东湖新技术开发区华工科技园	邮　编：430223
录　　排：华中科技大学惠友文印中心	
印　　刷：湖北新华印务有限公司	
开　　本：787mm×1092mm　1/16	
印　　张：8.75	
字　　数：292 千字	
版　　次：2023 年 5 月第 1 版第 1 次印刷	
定　　价：59.80 元	

本书若有印装质量问题，请向出版社营销中心调换
全国免费服务热线：400-6679-118　　竭诚为您服务
版权所有　侵权必究

Preface

Multimedia technology is a comprehensive product of both computer technology and social needs. In the early stages of computer development, people used computers to solve numerical computing problems in military and industrial production. With the development of computer technology, especially the development of hardware equipment, people began to use computers to process and represent images and graphics, so that computers can reflect natural things and calculation results in a more vivid way.

This textbook strives to combine basic theory and practical operation, focusing on cultivating the processing ability and material application ability of multimedia materials. There are five chapters and the main content includes the basic knowledge of multimedia technology, digital image processing, digital audio processing, digital video processing, and taking PowerPoint software as an example to introduce the related knowledge of multimedia platform software. This book is written by Li Ting, Lv Jianhong, Wang Yinlong, and translated by Liu Xiaofei, Wu Jie, Liu Ling, Zhang Ying, Peng Qi, and Wang Huan. Chapter One, Chapter Two, and Chapter Four are written by Li Ting. Chapter Three is written by Li Ting and Wang Yinlong. Chapter Five is written by Lv Jianhong. Chapter One is translated by Liu Ling. Chapter Two and Chapter Five are translated by Liu Xiaofei. Chapter Three and Chapter Four are translated by Wu Jie. Zhang Ying, Peng Qi, and Wang Huan are responsible for the amendment.

Grateful thanks to the leaders and professors who have provided valuable guidance and help in the process of compiling and translating this book.

This book is an application-oriented textbook that teaches profound knowledge in a simple way, focusing on application, and ability cultivation. Due to the limited knowledge of authors and translators, mistakes and flaws are inevitable in the compilation and translation. Readers are welcome to make comments and corrections.

Please scan the QR code to get supporting materials of this book.

Supporting materials

Textbook compilation group of *Multimedia Technology Basics*
September 2022

CONTENTS

Chapter 1 Preliminary Study on Multimedia Technology ······ (1)
  Section 1 Multimedia Technology and Multimedia System ······ (1)
    Ⅰ. Media ······ (1)
    Ⅱ. Multimedia ······ (2)
    Ⅲ. Multimedia Technology ······ (3)
    Ⅳ. Multimedia System ······ (4)
  Section 2 Basic Elements of Multimedia Information ······ (5)
    Ⅰ. Text ······ (5)
    Ⅱ. Graphic ······ (6)
    Ⅲ. Image ······ (6)
    Ⅳ. Audio ······ (7)
    Ⅴ. Video ······ (8)
    Ⅵ. Animation ······ (8)

Chapter 2 Digital Image Processing ······ (10)
  Section 1 Basic Knowledge of Digital Image ······ (10)
    Ⅰ. Digitalization of Image ······ (10)
    Ⅱ. Basic Properties of Image ······ (12)
    Ⅲ. Color Principle of Image ······ (13)
    Ⅳ. File Format of Image ······ (16)
  Section 2 Image Correction ······ (18)
    Ⅰ. Basic Editing Process of Digital Photos ······ (18)
    Ⅱ. Correction of Deformation and Secondary Composition ······ (18)
    Ⅲ. Adjust Size and Resolution ······ (23)
    Ⅳ. Correct Exposure ······ (24)
    Ⅴ. Adjust Color ······ (26)
    Ⅵ. Enhance Image Clarity ······ (28)
  Section 3 Image Beautification ······ (30)
    Ⅰ. Stamp Clone ······ (30)
    Ⅱ. Restore Texture ······ (32)
    Ⅲ. Beautify the Details ······ (35)
  Section 4 Establishment of Selected Area ······ (38)
    Ⅰ. Rectangular Marquee Tool Group ······ (38)
    Ⅱ. Lasso Tool Group ······ (40)
    Ⅲ. Quick Selection Tool Group ······ (42)
  Section 5 Multiple Layer Blending ······ (44)

	Ⅰ.	Draw Shapes	(45)
	Ⅱ.	Input Text	(50)
	Ⅲ.	Layer Blending	(51)
	Ⅳ.	Layer Style	(57)

Chapter 3 Operations of Digital Audio (61)

  Section 1 Fundamental Knowledge of Audio (61)
    Ⅰ. Fundamental Knowledge of Sound (61)
    Ⅱ. Digitization of Sound Signal (64)
    Ⅲ. Storage of Audio File (67)

  Section 2 Sound Recording and Noise Reduction (69)
    Ⅰ. Sound Recording (69)
    Ⅱ. Sound Editing (71)
    Ⅲ. The Noise Reduction of Audio File (72)

  Section 3 Sound Editing and Synthesis (75)
    Ⅰ. Adjustment of Sound (75)
    Ⅱ. Sound Synthesis (77)

  Section 4 Special Effect Processing of Sound (78)
    Ⅰ. Adding Echo (78)
    Ⅱ. Adding Reverb Effect (79)
    Ⅲ. Adjustment of Pitch (79)

Chapter 4 Digital Video Processing (81)

  Section 1 Fundamental Knowledge of Video (81)
    Ⅰ. Analog Video and Digital Video (81)
    Ⅱ. Linear Editing and Non-linear Editing (83)
    Ⅲ. Digitalization of Audio Signal (83)
    Ⅳ. Digital Video File Format (85)

  Section 2 Basic Editing Procedures of Video Footage (86)
    Ⅰ. Obtaining Footage (86)
    Ⅱ. Editing the Footage (89)
    Ⅲ. Adding Transition (93)

  Section 3 Operation of Special Effect of Video (95)
    Ⅰ. Adding Filter (95)
    Ⅱ. Picture-in-picture Effect (98)
    Ⅲ. Processing of "Blue Curtain" (100)

  Section 4 Adding Subtitles (103)

  Section 5 Saving and Output (106)
    Ⅰ. Saving (106)
    Ⅱ. Video Output (107)

Chapter 5 Application of Materials on PowerPoint of Multimedia Platform (110)

  Section 1 Introduction to Multimedia Platform Software (110)

	Ⅰ. Types of Multimedia Platform Software	(110)
	Ⅱ. Functions of Multimedia Platform Software	(111)
Section 2	Insertion and Editing of Materials	(112)
	Ⅰ. Image Material	(112)
	Ⅱ. Audio Material	(116)
	Ⅲ. Video Material	(119)
	Ⅳ. Comprehensive Applications of Materials	(121)
Section 3	Slide Design	(127)
	Ⅰ. Frame Design	(127)
	Ⅱ. Page Design	(128)
	Ⅲ. Design Tips for Beginners	(129)
References		(131)

Chapter 1 Preliminary Study on Multimedia Technology

Section 1　Multimedia Technology and Multimedia System

Ⅰ. Media

1. The Connotation of Media

Traditionally, media refers to the carrier of information expression and transmission, as well as the intermediary role between people to communicate and exchange ideas, thoughts or opinions, such as newspaper, radio, television, magazine, and so on in daily life. In computer science, media have two meanings: one is the physical entity that carries information, such as hard disc, compact disc, semiconductor memory, video tape, book, etc.; the other is the logical carrier of information, that is, the expression form of information, such as number, word, sound, graphic, image, video, and animation. The media in multimedia technology generally refers to the latter.

2. Classifications of Media

The development of modern science and technology has endowed media with many new connotations. According to the definition recommended by ITU-T, media is divided into the following five types.

(1) Perception Medium

Perception medium refers to a kind of media that can directly affect people's auditory, visual, tactile, and other senses, such as language, music, sound, graphic and image.

(2) Representation Medium

Representation medium refers to the intermediary medium that transmits perception medium. It is researched and constructed by man for processing, handling, and transmission of perception medium, namely the code used for data exchange. It is the expression of perception medium after digitization, such as voice and image coding, etc. The purpose of constructing representation media is to transmit perception medium more effectively from one side to the other for ease of processing and handling. There are various ways of encoding representation medim. For example, text can be compiled in ASCII code; audio can be encoded using PCM (pulse code modulation); static image can be encoded by still image compressing standard; moving image can be encoded by moving image compressing standard; video can be coded in different TV systems such as PAL, NTSC, and SECAM, etc.

(3) Presentation Medium

Presentation medium, also known as display medium, refers to a physical device that inputs perception media into a computer or displays through the computer, that is, computer input and output device that obtains and restores perception media, such as keyboard, camera, monitor, speakers, and so on.

(4) Storage Medium

Storage medium refers to the physical device that stores representation media, that is, the medium that stores the digitized code of the perception medium, such as soft disc, hard disc, compact disc, magnetic tape, paper, etc.

(5) Transmission Medium

Transmission medium refers to the physical carrier that transmits the representation medium, such as coaxial cable, optical fiber, twisted pair, and so on.

Among all types of media mentioned above, representation medium is the core.

Representation media can be divided into static medium and continuous medium. Static media is the representation of information, which has nothing to do with time, such as text, graphic, and image. Continuous medium has an implicit relationship with time, and its playback speed will affect the replay of the contained information, such as sound, animation, and video.

From the perspective of human-computer interaction, media can be divided into visual media, auditory media, tactile media, and so on. In human perception system, the information acquired by vision accounts for more than 60% of the total information; auditory information accounts for about 20% of the total information; touch, smell, taste, and so on are responsible for obtaining the rest of the information.

II. Multimedia

Multimedia is a form of information manifestation formed by the integration of more than one medium. It is the result of integrated processing and application of multiple media. To sum up, it is multi-media manifestation, multi-sensory functioning, multi-device supported, multi-disciplinary overlapping, and multi-field application.

The essence of multimedia is to digitize various forms of media information, and then use computers to process or handle digital media information, form an organic whole through logical links, and realize interactive control in a friendly way for users to use.

There are several differences between multimedia and traditional media: multimedia information is all digital signal, while traditional media information is basically analog signal;

traditional media can only make people passively receive information, while multimedia can make people actively interact with information; traditional media is generally in a single form, while multimedia is the organic integration of two or more different kinds of information.

Ⅲ. Multimedia Technology

1. The Connotation of Multimedia Technology

Usually, multimedia technology is linked with computer. It is an interactive processing technology that mainly based on computer technology, combined with other technologies like communication, microelectronics, laser, broadcast, television and other media information to enable integrated processing of multimedia information. To be specific, multimedia technology takes computer (or microprocessing chip) as the center, and integrates text, graphic, image, audio, video, animation and other information for digital comprehensive processing, making multiple media information logically-linked and interactive.

"Comprehensive processing" here mainly refers to the collection, compression, storage, control, editing, transformation, decompression, playback, and transmission of information, etc. In applications, multimedia generally refers to multimedia technology.

2. Characteristics of Multimedia Technology

From the perspective of research and development, multimedia technology has five basic characteristics: diversity, integratability, interactivity, real-time capability, and digitalization, which are also the five basic problems to be solved by multimedia technology.

(1) Diversity

Diversity refers to the diversity of media types and their processing technologies.

Multimedia technology involves a variety of information and information carriers. A variety of information carriers make information exchange in a more flexible way and a wider space. Diversity covers the following two aspects.

The first one refers to the diversity of information media. Various information carriers include hard disc, compact disc, voice, graphic, image, video, etc. The ability of computers to process and reproduce diverse information without distortion remains to be improved.

On the other hand, diversity refers to that while computer is processing the input information, it not only receives or reproduces information, but also exchanges, combines and processes the text, graphic, and video according to people's thoughts, so as to enrich the expression of artistic creation to achieve vividness, flexibility, and naturalness.

Diversity refers not only to the input of various information, that is, the capture of information, but also the output of information, that is, the presentation. The input and output are not necessarily the same. If the input is the same as the output, it is called a record or replay. If the input is processed, combined, and transformed, it is called authoring, which can better represent information, enrich its expressiveness, and enable users to receive information more accurately and vividly. This form has been widely used in film and television production in the past, and now it is also used in multimedia technology.

(2) Integratability

Integratability is mainly manifested in two aspects, namely the integration of multiple information media and the integration of software and hardware technologies as well as equipments and systems for processing these media. In multimedia system, all kinds of information media are not collected and processed in a single way as in the past, but are collected, stored, and processed in a unified manner by multiple channels at the same time, which emphasizes more on the cooperative relationship between all kinds of media and the utilization of a large amount of information contained in them. In terms of hardware, multimedia hardware system(including high-speed and parallel CPU, multi-channel input and output interface, broadband communication network interface, and large capacity memory, etc.)integrates all hardware devices into a unified system. In software, multimedia operating system manages multimedia application software and authoring tools software. The hardware and software of multimedia system are integrated into an information system capable of handling a variety of composite information media with the support of network.

(3) Interactivity

Interactivity refers to the effective control and use of information through various means, so that all parties involved(whether the sender or the receiver)can conduct editing, control and transmission. In addition to the free control in operation(through keyboard, mouse, touch screen, etc.), the integrated processing of media can also be done at will. When people completely enter into a virtual information world integrated with the information environment, a full range of interaction will enable people to experience realistic effects, which is the advanced stage of interactive application and this technology is called virtual reality technology.

(4) Real-time Capability

Since sound and video are time-dependent continuous media, thus multimedia technology must support real-time processing.

(5) Digitalization

The key device for processing multimedia information is the computer, so information in different media forms must be digitized, that is, data presented by a series of binary digits(0, 1).

All kinds of media information must be converted into digital form, before computer can store, process, control, edit, exchange, query, and retrieve them. Therefore, multimedia information must be digital. Digital media composed of bitstream are spread through computer and network. It changes the traditional relationship between information disseminator and audience, meanwhile it also changes the composition, structure and information dissemination process and effect.

IV. Multimedia System

Software, multimedia service system, and related multimedia data constitute an organic whole. Multimedia system is a kind of multi-dimensional information processing system that tends to be humanized. It takes the computer system as the core, uses the multimedia technology to realize the multimedia information (including text, sound, graphic, image, video, animation, etc.) collection, data compression and coding, real-time processing, storage, transmission, decompression, restoration output, and other comprehensive processing functions, and provides friendly man-machine interaction.

With the rapid development of computer network technology and multimedia technology, multimedia system has gradually developed into a network multimedia system to obtain services and contact with the outside world through the network.

Because of the diversity of multimedia data, the original materials are often distributed in different spaces and time, which makes the establishment and management of distributed multimedia database and multimedia communication the key technologies of multimedia computer system.

Multimedia resources have some special properties, therefore, multimedia systems often need to involve some special technologies, such as computer representation and compression of multimedia, multimedia database management, multimedia logic description model, multimedia data storage technology, multimedia communication technology, and so on.

From the current development and application trend of multimedia system, it can be roughly divided into two categories: one is the development system with the dual functions of editing and playing, which is suitable for professionals to produce multimedia software products; the other is the multimedia application system facing the actual users.

Section 2 Basic Elements of Multimedia Information

At present, the basic forms of multimedia information in computers can be divided into text, graphic, image, audio, video, animation, etc. These basic forms of information are also called the basic elements of multimedia information.

I. Text

Text is the information form expressed by characters, numbers, and various symbols. It is the information medium most used in real life, mainly used to describe knowledge.

There are two main forms of text: formatted text and unformatted text. In a text file, if there is only text information and no other information about the format, it is called an unformatted text file or plain text file; and text files with a variety of text typesetting information and other format information, are called formatted text files. Text content is organized in a linear way and the processing of text information is the most basic information processing. Text can be produced in the text editing software. Text files edited in Word and other editing tools can be imported into the multimedia application design. Text can also be directly produced in graphics software or multimedia editing software.

II. Graphic

Graphic refers to a variety of regular graphics drawn by computer graphics software from point, line and plane to three-dimensional space, such as straight lines, rectangle, circle, polygon and other geometric figures that can be represented by angle, coordinates, and distance.

Only the algorithm that generates the graph and some feature points on the graph are recorded in the graph file, so it is also called vector graph. Software that generates graphics by reading these instructions and converting them to the shapes and colors displayed on the screen is often called a drawing program. When the computer restores the output, the adjacent feature points are connected with specific multi-segment straight lines to form a curve, if the curve is closed, coloring method can also be applied to fill the color. The biggest advantage of graphics is that the individual parts of the processing diagram can be controlled separately, such as moving, rotating, zooming in or out, and twisting without distortion. Different objects can also overlap each other on the screen and maintain their own characteristics, or get separated if necessary. Therefore, graphics are mainly used to represent wire-frame drawing, engineering drawing, calligraphy, and so on. Most CAD and 3D modeling software use vector graphics as the basic graphic storage format.

Common vector graphic formats are 3DS(for 3D modeling), DXF(for CAD), WMF(for desktop publishing), and so on. The key of graphic technology is the production and reproduction of graphics. Graphic only stores algorithms and feature points, so compared with the large amount of data in images, it occupies less storage space. However, it needs to be recalculated each time when displayed on the screen. In addition, the graphic quality is higher than image quality in the printout and amplification.

III. Image

The image here refers to the still image. It can be captured from the real world, or generated by a computer. The image is a bitmap composed of units of pixels. Each pixel is encoded in binary numbers to reflect the color and brightness of the pixel.

Graphic and image are two different concepts in multimedia. The main differences are as follows.

a. Different structural principles. The basic elements of graphic are primitives, such as lines, points and surfaces, etc. The basic element of image is the pixel. A bitmap image can be understood as a matrix composed of pixel points.

b. Different data recording modes. Graphics are stored as drawing functions, while images are stored as pixel position information, color information, and gray level information.

c. Different processing operations. Graphics are usually edited by the Draw program, resulting in vector graphics. Vector graphics and primitives can be transformed independently by moving, scaling, rotating, and twisting. The main parameters of a graph include describing the instructions and parameters of the primitive location, dimension and shape. Image aregenerally processed with image softwares(Paint Brush, Photoshop and so on), which are mainly used for regular processing and editing of bitmap files and corresponding palette files. The transformation can not be applied to part of the image. Since bitmap files take up a large amount of storage space, data compression is generally required. Graphics can adapt to different resolutions without distortion during scaling, but images will distort when they are shrunk. Thus we can see that the whole image is composed of many pixels.

d. Different processing display speed. The graphic display process is carried out according to the sequence of primitives. It uses special software to convert the instructions describing the graphics into shapes and colors on the screen. The production process takes a certain amount of time. The image is to resolve the object with a certain resolution and then present the information of each point in a digital way, thus it can be displayed directly and quickly on the screen.

e. Different expressive force. Graphics are used to describe objects whose outlines are not too complex and whose colors are not very rich, such as geometric graphics, engineering drawings, CAD, 3D modeling and so on. Images can represent objects with a large number of details(such as scenes with complex light and shade changes and colorful contours). For example, photos and drawings. Complex images can be processed through image software to obtain clearer images or produce special effects.

Ⅳ. Audio

Audio refers to the continuous change of sound wave signal in the range of 20Hz ~ 20kHz. There are three elements of sound: tone, timbre, and loudness. Tone is related to frequency, timbre is determined by overtones mixed with tone, and loudness is related to amplitude. In terms of use, the sound can be divided into three forms: voice, music, and synthetic sound effects, and it can also be divided into waveform audio and MIDI audio from the perspective of processing.

1. Waveform audio

Waveform audio is used to represent sound wave in digital form, that is, sound waves such as voice and music effects are sampled, quantified, and encoded by sound cards and other special equipment, then converted into digital form, compressed and stored, and then decoded and restored to the original sound wave form when used.

2. MIDI audio

MIDI refers to electronic instrument digital interface. It was first used on electronic instruments to record the player's performance for later replay, and only after the introduction of the sound card that supported MIDI synthesis did it officially become a digital audio format for computers. It is a digital instruction sequence audio format for recording musical notation and the way the notes are played. It only contains very small amount of data.

MIDI audio is different from waveform audio. It does not sample, quantify, or encode sound waves. Instead, it records the playing information of electronic instrument keyboard (including key name, strength, and duration, etc.) These messages are known as MIDI messages which is the digital description of the score. The MIDI file corresponding to a piece of music does not record any sound information, but simply contains a series of MIDI messages that produce the music. Only reading MIDI messages is needed to generate the required musical instrument sound waveform, which can be output after amplification.

By integrating audio signals into multimedia, you can achieve results that no other medium can achieve. Not only can it heighten the atmosphere, but it can also increase the vitality. Audio information enhances the understanding of information expressed in other types of media.

Ⅴ. Video

Video refers to the continuous active image signal obtained from video camera, video tape recorder, DVD player, television receiver and other image output devices, that is, a number of related image data played continuously to form a video. These video images make the multimedia application system more powerful and more exciting. However, since the output of the above video signals is mostly standard full-color TV signals, in order to input it into the computer, not only the video signal should be captured to realize its conversion from analog signal to digital signal, but also demands compression and fast decompression, as well as the corresponding hardware and software processing equipment to play with. At the same time, the process will inevitably be affected by television technology.

There are three main types of television, namely NTSC(525/60), PAL(625/50), and SECAM(625/50). The Numbers in parentheses are the number of lines and frequency displayed on the TV. When the computer digitizes the signal, quantization, compression, and storage must be completed within specified time(such as 1/30 second). Video files can be stored in AVI, MPG, MOV, etc.

For the operation and processing of dynamic video, in addition to the same action and animation during play, special effects can also be added, such as hard cut, fade in and out, copy, mirror, mosaic, kaleidoscope, and so on, to enhance the performance. This in the media belongs to the content of media presentation attributes.

Ⅵ. Animation

Animation is ananimation created by computer animation design software. It is a continuous picture with motion sensation produced by continuous playing of images. The continuous play of animation refers to both the continuity of time and the continuity of image content, that is, the content difference between the two adjacent images is not big. Animation compression and quick play is also an important problem to be solved by animation technology, and there are many ways to tackle it. There are two methods for computer animation design: one is modeling animation, and the other is frame animation. The former is to design each moving object separately, give each object some characteristics, such as size, shape and color, and then use these objects to form a complete frame. Every frame in modeling animation is composed of modeling elements such as graphics, sound, text, and palette. The meta representation and behavior of each frame in the animation are controlled by the scripts composed of a production table. Frame animation is a continuous picture made up of bitmaps. Just like cinematography or video, every screen is needed to be designed separately.

When the computer is making animation, as long as the active screen can be done, the rest of the middle screen can be completed by the computer interpolation. The non-moving parts can be directly copied, consistent with the active painting. When these images show only two-dimensional perspective, they are two-dimensional animation. If using the form of CAD to create an image of the space, it is a three-dimension animation. If we make it have real lighting effect and texture, it will become three-dimension reality animation. The file format of storing animation includes FLC, SWF, and so on.

The common feature of video and animation is that every image is contextual. Generally speaking, the back image is a transformation of the front image. Each image is called a frame. When frames are continuously projected on the screen, they will create a feeling of continuous motion. When the play speed is above 24 frames per second(FPS), there is a natural sense of continuity in vision.

Chapter 2

Digital Image Processing

Section 1 Basic Knowledge of Digital Image

Ⅰ. Digitalization of Image

In the real world, pictures, pictorials, drawings, and other image signals are continuous functions within space, in gray or color. In order to process images in computer, we must first digitize them and then analyze them with computer.

The process of image digitalization is divided into three steps: sampling, quantization and encoding.

1. Sampling

The discretization of an image in two-dimensional space is called sampling. The discrete points of an image after sampling are called sample points(pixels). The essence of sampling is to describe an image with several points. In short, it is to divide the spatial continuous image into rectangular mesh structure with equal space in horizontal and vertical directions. In this way, an image is sampled into a set of limited pixels, and the micro grid is called pixel point. For example, a 640×480 image means that the image is composed of 307200 pixels.

In the process of sampling, the selection of sampling interval is very important, which determines how good the sampled image can reflect the original image. The density of the sample determines the resolution of the image. Generally speaking, the more complex and colorful the original image is, the smaller the sampling interval should be, and the more sampling points are, the better the image quality will be.

2. Quantization

Discretization of the gray or color values of each pixel obtained after image sampling is called image quantization. Generally, L-bit binary number is used to describe the gray or color value, and the quantization series is 2^L. Generally, 8-bit, 16-bit or 24-bit can be used as quantization bits. Higher quantization bits mean that more available colors and more accurate color representation are available, which reflect the color of the original image better. However, the corresponding capacity of the digital image is larger.

For example, a gray-scale photograph is continuous in both horizontal and vertical directions. A discrete sample of $M \times N$ can be obtained by sampling at equal intervals along the horizontal and vertical directions. The value of each sample point represents the gray level (brightness) of the pixel. The gray level is quantized to make its value become a limited number of possible values. The image obtained by such sampling and quantization is called digital image. As long as there are enough horizontal and vertical sampling points and enough quantization bits, the quality of the digital image is no inferior to that of the original image.

The basic problem of both sampling number and quantization bit is the choice of image visual effect and storage space.

The digitized bitmap can be described by the following information matrix, and its elements are the gray level or color of pixels.

$$\begin{bmatrix} f(0,0) & f(0,1) & f(0,2) & \cdots & f(0,n-1) \\ f(1,0) & f(1,1) & f(1,2) & \cdots & f(1,n-1) \\ f(2,0) & f(2,1) & f(2,2) & \cdots & f(2,n-1) \\ \vdots & \vdots & \vdots & & \vdots \\ f(m-1,0) & f(m-1,1) & f(m-1,2) & \cdots & f(m-1,n-1) \end{bmatrix}$$

3. Encoding

There are two functions of encoding: one is to record digital image data with a certain format; the other is to adopt a certain coding technology to compress the data due to the huge amount of digital image data, so as to reduce the storage space and improve the image transmission efficiency.

Digital image compression is based on two facts. First, data redundancy is present in image data. For example, the color of adjacent sampling points in an image often has spatial coherence, while the color representation of pixel based on discrete pixel sampling usually does not make use of this spatial coherence, resulting in spatial redundancy. Second, people are not sensitive to vision. For example, "visual masking effect" is working on human eyes, that is, people are sensitive to brightness, but not sensitive to sharp changes in edges, and the resolution of color details is much lower than that of luminance details. When recording the original image data, it is usually assumed that the visual system is linear and uniform, and the sensitive and insensitive parts are equally processed, thus generating more data than the ideal encoding (i. e. the encoding that distinguishes the sensitive and insensitive parts).

At present, many developed encoding algorithms have been applied to image compression. Common image encoding includes run length encoding, Huffman encoding, LZW encoding, prediction encoding, transform encoding, wavelet encoding, artificial neural network, etc.

Since the 1990s, a series of international standards for static and dynamic image coding have been developed and are being developed by ITU, ISO and IEC. The approved standards include JPEG standard, MPEG standard and H. 261 standard. These standards and suggestions are the summary of the achievements and experiences of the experts from various countries working in the corresponding fields. Because of these international standards, the image encoding, especially the video image encoding and compression technology are developing rapidly. At present, a large number of hardware, software products and ASIC based on these standards have emerged in the market(such as image scanners, digital cameras, digital cameras, etc.). It plays an important role in the rapid development of modern image communication and the development of new application fields of image encoding.

II. Basic Properties of Image

Bitmap is composed of pixels, the density of pixels and the color information of pixels directly affect the image quality. The image property is adopted for describing an image. Image attributes include resolution, pixel depth, true/false color, etc.

1. Resolution

Common resolutions include image resolution, display resolution and print resolution.

(1) Image Resolution

Digital image is composed of a certain number of pixels. Image resolution refers to the pixel density of an image, that is, the number of pixels per inch, which is expressed in PPI (pixels per inch). For an image of the same size, the higher the resolution of the image, the more pixels constitute the image, and the clearer the image details. On the contrary, the coarser the image appears. The image effects of different resolutions are shown in Figure 2-1.

Figure 2-1 Effect of Image Resolution on Vision

(2) Display Resolution

Display resolution refers to the number of pixels that can be displayed on the display screen. For example, a display resolution of 1024 × 768 means that the display screen is divided into 768 lines (vertical resolution). Each line displays 1024 pixels (horizontal resolution), and the whole display screen contains 786432 imaging points.

The more pixels the screen can display, the higher the resolution of the display device, the higher the image quality. On the display screen with a fixed size, the higher the display resolution, the finer the display image, but the smaller the picture will be.

(3) Print Resolution

Print resolution is the number of ink spots that can be printed on an inch of paper, expressed in DPI(dots per inch). The resolution of the printing device is between 360 dpi and 2400 dpi. The larger the resolution, the smaller the ink dot of the image output(the size of the ink dot is only related to the hardware technology of the printer, independent of the resolution of the image to be output), and the better the output image effect.

2. Pixel Depth

Pixel depth, also known as color depth, refers to the number of bits used to store the color(or gray-scale)of each pixel. Pixel depth determines the maximum number of colors that can be used in a color image, or determines the gray level of a gray-scale image. Larger pixel depth means that digital images have more available colors and more accurate color representation. For example, for a color image in RGB mode, 8 bits are used for each R, G and B components, which means that the depth of the pixel is 24. Each pixel can be one of the 2^{24} =16777216 colors. The gray image effects of different pixel depth are shown in Figure 2-2.

Gray-scale-256

Gray-scale-16

Gray-scale-4

Figure 2-2　Effect of Pixel Depth on Image Vision

III. Color Principle of Image

The sense of color is the most popular form of general aesthetic feeling. Human being is very sensitive to color. Image is closely related with color, and on the other hand, color is an important part of the image.

1. Tricolor

The basic color that can't be decomposed is called primary color. Primary color can synthesize other colors, but other colors can't restore to primary color. As shown in Figure 2-3, red, green and blue can be mixed in certain proportion to obtain other colors, and any one of the three cannot be generated by mixing the other two colors. The three independent colors are called tricolor in chromatics.

The color of TV picture tube, LED display screen and other display images are composed of red, green and blue lights, which are also known as tricolor.

In art, magenta with a small amount of yellow can generate bright red, but scarlet can not produce magenta; green and a small amount of magenta can get blue, while blue and white make dull green. Therefore, magenta, cyan and yellow are the three primary colors in color printing, such as ink mixing, principle and production of color photo, design and practical application of color printer. They are often called pigment or printing primary colors, as shown in Figure 2-4.

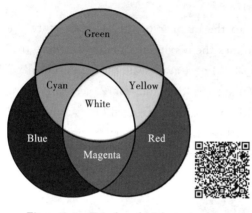

Figure 2-3 Tricolor of Light   Figure 2-4 Tricolor of Pigment

2. Principle of Color Mixing

According to the tricolor principle, any one color(except the three primary colors)can be mixed in different proportions through the three primary colors. Color mixing can be divided into additive mixing and subtractive mixing. Color mixing can occur after entering the vision, which is called neutral mixing.

(1) Additive Mixing

Additive mixing refers to the mixing of colored lights, and the color mixing relationship is shown in Figure 2-3. When two or more kinds of light are mixed together, the brightness will be improved. The total brightness of the mixed color light is equal to the sum of the brightness of the mixed colors. If white light can be generated only by mixing two colored lights, the two lights are complementary colors. For example, red + cyan = white, then red and cyan are complementary colors to each other.

(2) Subtractive Mixing

Subtractive mixing mainly refers to the mixing of pigments, and the color mixing relationship is shown in Figure 2-4. Subtractive mixing takes advantage of the filtering property, that is, subtracting unwanted colors from white light and leaving the desired colors. If two colors can produce gray or black, they are complementary colors. For example, red + cyan = black, then red and cyan are complementary colors. When the three primary colors are mixed in a certain proportion, the color obtained can be black or black gray. In subtractive mixing, the more colors are mixed, the lower the lightness and the lower the purity.

(3) Neutral Mixing

Neutral mixing is a kind of visual color mixing based on human visual physiological characteristics. It does not change the color light or luminous material itself, and the brightness of color mixing effect does not increase or decrease. The common methods are color disc rotation blending and spatial vision blending.

3. Color Mode

Color mode is a model that represents a certain color as a digital form, or a way to record image color. Common color modes are: RGB, CMYK, HSB, LAB, gray mode, bitmap mode, etc.

(1) RGB Mode

RGB mode corresponds to computer monitor, TV screen and other display devices, so it is also called color light mode. When R, G and B are all zero, that is, the three colors are not luminous at the same time, it becomes black. The stronger the light, the brighter the color. When R, G, and B are 255, that is to say, when all three colors emit at the same time, they will become white. Therefore, RGB mode is called additive color method.

(2) CMYK Mode

CMYK mode, also known as printing color mode, is the standard color for printing, which is mainly used in printers. The four letters refer to cyan, magenta, yellow and black (represented by the last letter K). In printing, they represent four colors of ink. There is no difference between CMYK mode and RGB mode in essence, but the principle of color generation is different. In RGB mode, the color light emitted by the light source is mixed to generate color; in CMYK mode, the light is illuminated on the paper with different proportions of C, M, Y, K ink, and part of the spectrum is absorbed, and the light reflected to the human eye produces color. When C, M, Y and K are mixed into color, with the increase of C, M, Y and K, the light reflected to the human eye will be less and less, and the brightness of the light will be lower and lower. Therefore, the method of color generation in CMYK mode is also called chromatic subtraction.

(3) HSB Mode

HSB is a color mode based on human visual system(color). In other words, HSB mode is a color mode based on the visual characteristics of human eyes, which is more in line with the law of human eye to observe color than RGB mode. If we observe a color, such as the color of the Army uniform shirt, we will not say what its RGB value is, but say it is green, to be more accurately, light green. According to the visual characteristics of human eyes, HSB mode divides color into three factors: hue, saturation and brightness.

H is the hue(degree), which is the appearance of the color. It indicates the color of the object reflected into our eyes. It is determined by the wavelength of the light. It is used to adjust the color in the model, and the value is $0 \sim 360$ degrees.

S is the saturation(%), also known as purity, refers to the depth or brightness of the color. Pure spectral color, that is, the color of rainbow is fully saturated. Other colors, such as pink and light green, can be regarded as pure spectral color light mixed with white light. The proportion of pure spectrum color light is its saturation, and its value is 0(gray)~100%(pure color).

B is the brightness(%), also known as lightness. It refers to the intensity of the light we see. The value is 0(black)~ 100%(white).

(4) LAB Mode

LAB mode was published by CIE in 1976, which theoretically includes all the color patterns that can be seen by human eyes. LAB color is converted from RGB three primary colors, which is the bridge from RGB mode to HSB mode and CMYK mode. LAB mode does not depend on light or pigment, which makes up for the deficiency of RGB and CMYK color modes. It is the color mode used by Photoshop when converting different color modes. Users can use LAB mode in image editing. Furthermore, when RGB mode is converted to CMYK mode, it is accompanied with color loss which is not present in the transformation between LAB mode and CMYK mode. Therefore, the best way to avoid color loss is to edit the image in LAB mode and then convert it to CMYK mode for printing. However, some Photoshop filters do not work on LAB mode images, so if users want to process color images, it is recommended to choose either RGB mode or LAB mode. Before printing and outputting, convert into CMYK mode, and there is no need for color correction to convert images in LAB mode.

(5) Gray Mode

If the gray mode is selected, there is no color information in the image, the color saturation is 0, and the image has 256 gray levels, ranging from 0(black) to 255(white). If users want to edit and process black-and-white images, or convert color images to black-and-white images, set the mode of the image as gray-scale. Since the color information of gray-scale images is removed from the file, the file size of gray-scale image is much smaller than that of color.

(6) Bitmap Mode

Black and white are adopted to represent pixels in the image. Bitmap mode image is also called black and white image, because there are only two colors in the image. Except for special purpose, this mode is not selected. When the color mode needs to be converted to bitmap mode, it must be converted to gray mode first, and then the gray mode can be converted to bitmap mode.

(7) Index Mode

Index mode uses 256 colors to represent images. When an RGB or CMYK image is converted to index mode, a 256 colors table is established to store the colors used in the image. Therefore, the image of index mode occupies less storage space, but the image quality is not high, which is suitable for multimedia animation and web image production.

Ⅳ. File Format of Image

The image file format is the embodiment of the image format stored in computer and the data compression coding method. Different file formats are distinguished by different file extensions. Generally, the recognition and application of image files, even the conversion between file formats can be realized by image processing software. At present, there are many common formats, such as BMP, JPEG, GIF, TIFF, PNG, and so on. The default image file of Photoshop is PSD format. Since most image formats do not support the properties of Photoshop such as layers, channels and vector elements, if users want to continue editing the image, it is necessary to save the image in PSD format.

1. BMP Format

BMP(full name: Bitmap) is a bitmap format file developed by Microsoft for Windows. It is independent of hardware devices and is widely used. It uses bit mapping storage format, and does not use any other compression except image depth. Therefore, BMP file takes up a lot of space. The image depth of BMP file can be 1 bit, 4 bit, 8 bit and 24 bit. When BMP files store data, the scanning mode of images is from left to right and from bottom to top.

2. JPEG Format

JPEG(Joint Photographic Expert Group) is the product of JPEG standard, which is formulated by the International Organization for Standardization (ISO). It is a lossy compression standard for continuous tone still images.

JPEG format can compress the image in a small storage space, and the repeated or unimportant data in the image will be lost, so it is easy to cause the loss of image data. If the compression ratio is too high, the quality of the decompressed image will be significantly reduced. Therefore, if high-quality image is expected, high compression ratio is not suggested. However, JPEG compression technology is very advanced. It uses lossy compression to remove redundant image data. It can obtain high compression rate and display very vivid image. In other words, it can get better image quality with the smallest storage space. Moreover, JPEG is a very flexible format, which has the function of adjusting image quality. It allows the file to be compressed with different compression ratios. It supports multiple compression levels. The compression ratio is usually between 10 : 1 and 40 : 1. The larger the compression ratio, the lower the quality; on the contrary, the smaller the compression ratio, the higher the quality. JPEG format mainly compresses high-frequency information and retains color information well. It is suitable for Internet and can reduce image transmission time. It can also support 24bit true color and is widely used in images requiring continuous tone. Therefore, it is the most popular image format on the network.

3. GIF Format

GIF(Graphics Interchange Format) is an image file format developed by CompuServe company in 1987. GIF file data is a lossless compression format of continuous tone based on LZW algorithm. Its compression rate is generally about 50%. It does not belong to any application program. At present, almost all relevant softwares support it. GIF image file is widely used in software of public domain.

4. TIFF Format

TIFF(Tag Image File Format) file is a common image file format developed by Aldus and Microsoft Company for desktop publishing system. TIFF format is flexible and changeable. It defines four different formats: TIFF-B is suitable for binary images; TIFF-G is suitable for black-and-white gray-scale images; TIFF-P is suitable for color images with color palette; and TIFF-R is suitable for RGB true color images.

5. PNG Format

PNG(Portable Network Graphics) is the latest image file format commonly used on the Internet. PNG can provide lossless compressed image files which are 30% smaller than GIF. It also provides 24bit and 48bit true color image support and many other technical support. PNG images support storing images as transparent background. Since PNG is relatively new, it is not applicable for all the programs to store image files. But Photoshop can process PNG image files, and can also store them in PNG image file format.

6. PSD Format

PSD(Photoshop Document) is a special file format of Photoshop image processing software. The file extension is". PSD", which can support various image features of layers, channels, masks and different color modes. It is a non-compressed saving format of original file. The scanner cannot directly generate files in this format. PSD file sometimes has a large capacity, but it is the best choice to save the unfinished image as PSD format in image processing, for it can retain all the original information.

Section 2　Image Correction

Ⅰ. Basic Editing Process of Digital Photos

Although the function of digital camera is more and more powerful, the quality of the photos is not ideal due to the influence of environment, the defect of camera itself, or the bad composition of picture… However, it is not necessary to abandon these photos which can be recovered into a new look after some renovation with Photoshop.

First of all, we will understand the basic workflow of digital photo editing, and then we will learn the specific methods of digital photo editing. The basic workflow of digital photo editing can be summarized into five steps as shown in Figure 2-5.

However, it needs to be clear that the problems of each photo will not be exactly the same. Some only need to adjust the exposure, but the color has no problem; some may have skew, too dark, and color deviation at the same time. When editing photos, we should check one by one according to the above five steps, and adjust the contents that need to be adjusted, instead of all the contents mentioned above.

Ⅱ. Correction of Deformation and Secondary Composition

To edit a photo, we should first check whether there is any deformation or other missing composition in the image, such as the horizontal line skew, obvious deformation of buildings, and the appearance of objects cut into half of the boundary. The purpose is to determine the content of the photo, so that when adjusting the brightness, contrast, or color of the image,

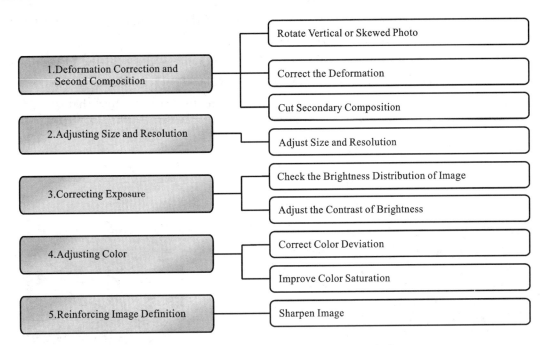

Figure 2-5 Operating Process of Editing Digital Photos

appropriate correction can be made according to the correct document.

1. Rotate Vertical or Skewed Photo

(1) Rotation of Photo

For the photo shown in Figure 2-6, it needs to be rotated 90 degrees anticlockwise. Specific operation method: click the 【Image】 menu, select the 【Image Rotation】 menu item, and select 【90° CCW】 in the cascade menu, as shown in Figure 2-7, to complete the image straightening.

Figure 2-6 Vertical Photo

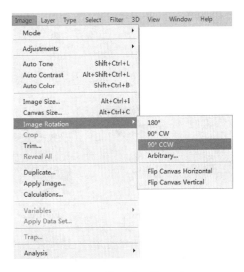

Figure 2-7 Rotating Procedures

(2) Correction of Tilt

As shown in Figure 2-8, it is necessary to determine the tilt angle of the photo before rotating it. The specific operation steps are as follows:

Step 1. Find the 【Eyedropper Tool】 in the 【Toolbox】, press and hold the 【Eyedropper Tool】 button to open the tool group list, select the 【Ruler Tool】, as shown in Figure 2-9, drag along the horizontal plane or horizon on the photo to draw the line that should be horizontal, as shown in Figure 2-10.

Figure 2-8　Tilted Photo　　　Figure 2-9　Ruler Tool　　　Figure 2-10　Draw Horizontal Line with Ruler Tool

Step 2. Click the 【Image】 menu, select the 【Image Rotation】 menu item, and select 【Arbitrary】 in the cascade menu, as shown in Figure 2-11. The default value in the pop-up dialog box is calculated by the computer through the straight line drawn by the ruler, as shown in Figure 2-12. Do not change, just confirm directly. The rotated picture is shown in Figure 2-13.

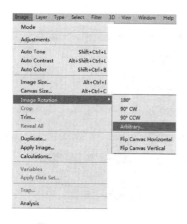

Figure 2-11　Arbitrary Option　　　Figure 2-12　Set Rotating Angle　　　Figure 2-13　After Rotating

(3) Cut Secondary Composition

After the image is rotated, the edge of the image is inclined, so it is necessary to crop it to make a rectangular photo. The operation steps are as follows:

Step 1. In the 【Toolbox】, select the 【Rectangular Marquee Tool】, as shown in Figure 2-

14. Set the style as a 【Fixed Ratio】 in the property bar of the 【Rectangular Marquee Tool】, with the aspect ratio of 3 : 2, as shown in Figure 2-15.

Step 2. Drag the selection box from the image, as shown in Figure 2-16. In this process, press the left mouse button and hold down the space bar at the same time so that the position of the selection box can be adjusted. After releasing the space bar, users can continue to adjust the size of the selection box.

Step 3. Click the 【Image】 menu and select the 【Crop】 menu item, as shown in Figure 2-17, so as to obtain the trimmed image, as shown in Figure 2-18. Finally, press Ctrl + D to cancel the selection box.

2. Correct the Deformation

For photos with lens distortion caused by too large elevation angle, as shown in Figure 2-19, it is necessary to correct the lens distortion first and then cut it. The specific operation steps are as follows:

Figure 2-14 Rectangular Marquee Tool

Figure 2-15 Property Bar of Rectangular Marquee Tool

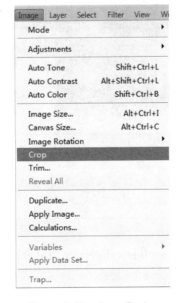

Figure 2-16 Drag a Rectangular Box

Figure 2-17 Crop Option

Figure 2-18 After Cropping

Step 1. Click the 【Filter】 menu, select the 【Lens Correction】 menu item, as shown in Figure 2-20.

Step 2. Click the 【Move Grid Tool】 button on the left of the pop-up menu so as to show

Figure 2-19 Photo with Distorted Lens   Figure 2-20 Lens Correction Option

reference grid, then click the Custom on the right part. Adjust the parameters as shown in Figure 2-21, adjust the Vertical Perspective to -28 and the Horizontal Perspective to 10.

Figure 2-21 Parameters Setting of Lens Correction

Step 3. Cut the photo into a rectangle. The clipping method is shown in the section of Cut Secondary Composition.

Tips:

1. In the Lens Correction window, the gray grid line on the image is the reference line, which can help us to adjust the image.

2. Description of Transform option group:

Adjust the Vertical Perspective. If it is set to a negative value, the lower part will shrink and deform, and if it is set to a positive value, the upper part will shrink and deform;

Adjust the Horizontal Perspective. If it is set to a negative value, the right side will shrink and deform, and if it is set to a positive value, the left side will shrink and deform;

Adjust the Angle and rotate the image with the center;

The setting way to fill the edge space after distortion: find Auto Scale Image, Edge extension: repeated image edge pixels; Transparency: transparent background; Black color: fill in the background with black; White color: fill in the background with white.

Ⅲ. Adjust Size and Resolution

With the improving functions of digital cameras, the photos taken are generally millions or even tens of millions of pixels. Although such photos are very clear, they take up too much storage space, so generally we do not need such large photos. Therefore, after taking photos, it's necessary to adjust the size and resolution of photos according to the purpose of the photos. The adjustment method is as follows:

Step 1. Click the 【Image】 menu and select the 【Image Size】 menu item, as shown in Figure 2-22.

Step 2. In the open window as shown in Figure 2-23, set the Image Size and Resolution parameters as shown in Figure 2-24, and confirm to complete the adjustment.

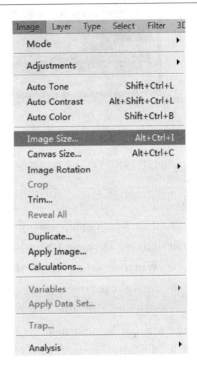

Figure 2-22 Image Size Option

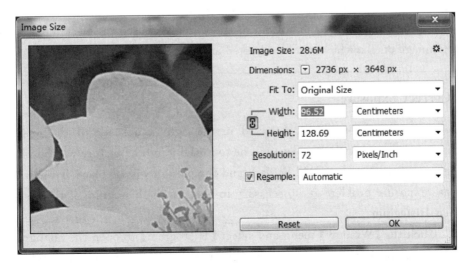

Figure 2-23 Image Size Dialog Box

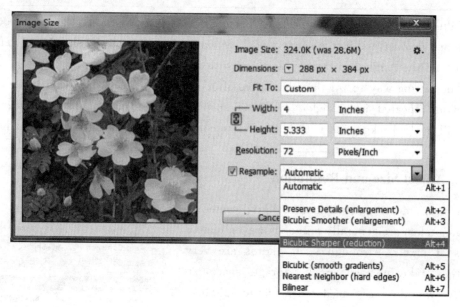

Figure 2-24　Image Parameters after Adjusting

Tips:

1. The relationship among pixel dimensions, document size and resolution is: Pixel Dimensions = Document Size × Resolution.

2. When adjusting, check the Constraint Proportions check box. When adjusting the width or height parameters, adjust one of them, and the other will change automatically. This can ensure that the length width ratio of the image will not change, and then the image will not be deformed.

3. The Resample Image check box is used to determine whether to change the number of pixels in the image. If it is selected, the number of pixels will be changed according to the size and resolution of the document.

4. The drop-down list at the bottom of the window is used to select the adjustment method of image pixels, which can be adjusted according to the instructions in brackets.

Ⅳ. Correct Exposure

The exposure of the image is reflected in the light and shade of the image. If the exposure is too strong, the whole image will be too bright and the picture will be white; if the exposure is too weak, the image will be dark and the picture will be black. To judge whether an image has exposure problem, it is necessary to observe the histogram of the image.

1. View Histogram

Step 1. Click the 【Window】 menu and select 【Histogram】 to open the histogram panel.

Step 2. Click the button in the upper right corner of the histogram panel and select 【Expanded View】 in the opening cascade menu, as shown in Figure 2-25.

Step 3. In the histogram expanded view panel, select the RGB option in the Channel

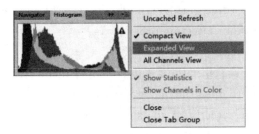

Figure 2-25 Expanded View Option

drop-down list, as shown in Figure 2-26, and open the brightness histogram of the image, as shown in Figure 2-27.

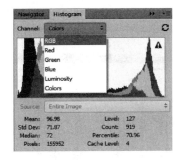

Figure 2-26 RGB Option

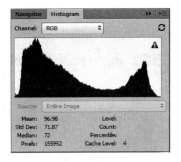

Figure 2-27 Brightness Histogram

2. Automatic Adjustment

Click the 【Image】 menu and select 【Auto Tone】 or 【Auto Contrast】 or 【Auto Color】 to realize the automatic adjustment of the image.

The automatic adjustment can be completed quickly, but it does not provide any adjustment parameter setting, so the adjustment result is unpredictable. Therefore, it is generally not recommended.

3. Manual Adjustment

Photoshop provides us with three tools to manually adjust the exposure of an image, and the adjustment ranges are different.

(1) Whole Adjustment

Click the 【Image】 menu, select the 【Adjustments】 item, and select the 【Brightness /Contrast】 option in the cascade menu to open the dialog box, as shown in Figure 2-28. By adjusting the position of each slider in the dialog box, the entire image can be adjusted. Adjusting the brightness can make the image bright or dark, and adjusting the contrast can increase or decrease the contrast degree. However, this method has many limitations. It can't satisfy the need that part of the image is expected to be modified.

Figure 2-28 Brightness/Contrast Dialogue Box

(2) Partial Adjustment

Click the 【Image】 menu, select the 【Adjustments】 menu item, and select the 【Levels】 option in the cascade menu to open the dialog box, as shown in Figure 2-29. This function divides the image level into three parts: highlights, midtones and shadows. Each part can be adjusted independently without affecting other parts.

(3) Individual Adjustment

Click the 【Image】 menu, select the 【Adjustments】 menu item, and select the 【Curves】 option in the cascade menu to open the dialog box as shown in Figure 2-30. In the dialog box, drag the input-output curve directly to adjust the direct corresponding relationship between the input level and the output level, so as to achieve the purpose of adjusting the image brightness. This function can be adjusted for 256 color gamuts respectively, with maximum flexibility. The button marked in red circle is the image adjustment tool, which allows us to adjust the curve directly on the image. After clicking the button, put the mouse at the place to be adjusted. Drag up to improve the brightness of the color level, and drag down to reduce the brightness of the color level.

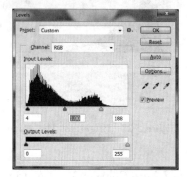
Figure 2-29　Levels Dialogue Box

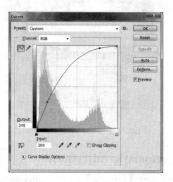
Figure 2-30　Curves Dialogue Box

Tips:

1. The histogram shown in Figure 2-27 is the histogram of an image with ideal exposure. When the exposure is corrected, it is necessary to observe the change of the histogram, and try to make the histogram spread over all levels, and there is no obvious "peak" or "valley" on the histogram.

2. The histogram of the image is not absolutely evenly distributed, so it should be judged according to the content of the picture. For example, when taking a picture with snow scenery, it is normal that the whole picture is white and should not be over exposed.

3. When using the Image Adjustment tool in the [Curves] dialog box to adjust the image, although it is dragged in a part of the image, all the pixels with the same color gamut on the image are adjusted.

Ⅴ. Adjust Color

When taking photos, photo color might be distorted due to the influence of illumination and surrounding scenery. For example, when taking pictures of sunrise, the picture tends to be red.

1. Correct Color Deviation

To judge whether there is color deviation in an image, it is generally necessary to observe the color areas such as black, white and gray in the picture. If the values of R, G and B color components in these areas are quite different, it can be considered that color deviation is present in the image. Among the three color components, which component value is the largest, the image is biased to which color. Therefore, in the process of image color deviation correction, it is necessary to determine whether there is color deviation in the image first, and then correct it. The specific steps are as follows:

(1) Judge Color Deviation

Step 1. Click the 【Window】 menu, select the 【Info】 item, and open the information panel, as shown in Figure 2-31.

Step 2. Select the 【Eyedropper Tool】 in the 【Toolbox】. When the mouse is over the image, the color value of the point will be displayed on the information panel. Users can also press and hold the Shift key to select the sample points(4 at most)in the neutral color areas such as black, white and gray, as shown in Figure 2-32. Observe whether the R, G and B values are close on the information panel, as shown in Figure 2-33. If they are not close, it indicates that color deviation is present in the image, which needs to be further adjusted.

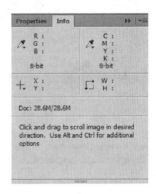

Figure 2-31 Information Panel

Figure 2-32 Sample on the Image

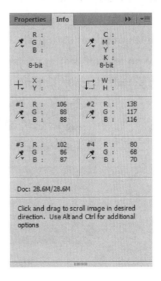

Figure 2-33 Information Panel after Sampling

(2) Correct Color Deviation

Click the 【Image】 menu, select the 【Adjustments】 item, and select the 【Photo Filter】 option in the cascade menu to open the dialog box as shown in Figure 2-34. In accordance with the complementary color principle, directly reduce the biased color components or increase their complementary colors, so that the values of the three color components are basically close, as shown in Figure 2-35. The adjusted values of each color component are shown in the following column.

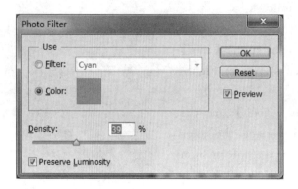

Figure 2-34 Photo Filter Dialog Box

Figure 2-35 Information Panel after Adjusting

2. Improve Color Saturation

Click the 【Image】 menu, select the 【Adjustments】 menu item, and select the 【Vibrance】 option in the cascade menu to open the Vibrance dialog box as shown in Figure 2-36. The function of this option is to adjust the entire color. The Vibrance slider will adjust all colors by the same amount; adjusting the saturation slider will distinguish the current saturation status of the color, and only increase the saturation for the color with low saturation, so as to avoid losing details due to clipping the color approached to saturation.

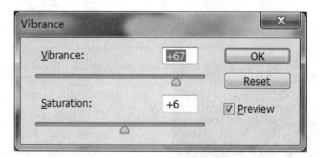

Figure 2-36 Vibrance Dialog Box

Ⅵ. Enhance Image Clarity

The purpose of enhancing the image clarity is to make the details on the image clearer. The details on the image are the places where the color changes violently in the picture. Therefore, the image sharpness is enhanced by sharpening the image. The specific operation steps are as follows:

Click the 【Filter】 menu, select 【Sharpen】, and then select 【Unsharp Mask】 in the cascade menu to open the dialog box, as shown in Figure 2-37. The function of image sharpening is to make the image clearer, and its essence is to make the color difference greater where the color changes violently.

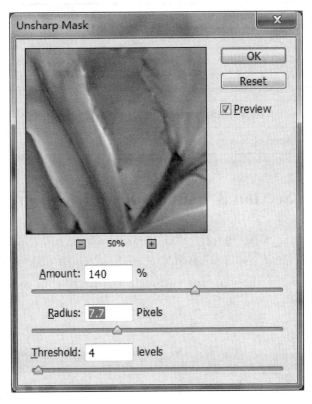

Figure 2-37　Unsharp Mask Dialog Box

The upper left corner of the Unsharp Mask window is the part of the image to be processed. The display scale can be adjusted by "＋" and "－" below it. The default value is 100%. After adjusting the parameters, the upper left corner displays the adjusted picture effect. Move the mouse to the local screen, press and hold the left mouse button to see the image effect before adjustment. Releasing the left mouse button, the adjusted image effect is displayed. The sharpening effect of the image can be compared by pressing and releasing the left mouse button.

Here are the description of parameters. Amount: the degree to enhance the contrast of image edge pixels; Radius: pixels quantity affected by sharpening at the edge of the image; Threshold: the critical value between two pixels differences that are considered to be the image edge.

Generally, the initial value of quantity and threshold is set first. If the quantity is set between 100% and 300%, the threshold value is set as 0, so that the image remains the most sensitive state; then, adjust the radius. For images with obvious outline, such as machines and buildings, a higher radius can be used, such as 1 pixel～2 pixels. For images with delicate and soft outlines such as people and plants, the radius should be lower, about 0.5 pixel~1 pixel; then adjust the number and threshold as needed.

Unsharp Mask parameter setting can refer to Table 2-1.

Table 2-1 Reference Value of Unsharp Mask Parameters

Parameters Image Contents	Amount	Radius	Threshold
gentle figures	80%~120%	1~2	10
scenery, flower, animal	100%~200%	0.7~1	3~5
building, machinery, and car with obvious outlines	80%~150%	1~4	5~10
general photos	120%	1	4

Section 3 Image Beautification

Sometimes there are some defects that damage the picture, such as messy wires or sundries that are difficult to avoid when taking photos, which makes the picture look very messy. Don't rush to delete these photos with defects. After being processed by Photoshop, all the defects might be cleaned up.

Ⅰ. Stamp Clone

Clone stamp tool is used to cover the partial content of the image to other parts of the image. It is suitable for repairing defects or copying objects in the image. The specific operation steps are as follows:

Step 1. Click 【Create a new layer】 at the bottom of the Layers Panel, as shown in Figure 2-38, create a blank layer above the background layer, and select the blank layer, as shown in Figure 2-39.

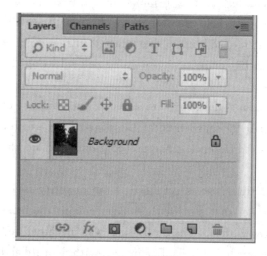
Figure 2-38 Layers Panel

Figure 2-39 Newly-established Blank Layer

Step 2. Select 【Clone Stamp Tool】 in Toolbox, as shown in Figure 2-40.

Step 3. Set the Mode, Opacity, Flow and Sample source in the property bar of the Clone Stamp Tool, as shown in Figure 2-41.

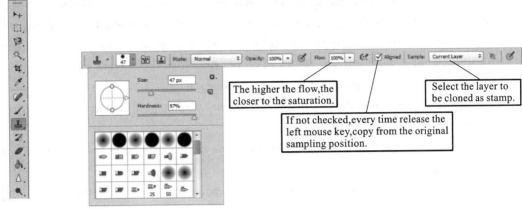

ure 2-40　Clone Stamp Tool

Figure 2-41　Property Setting of Clone Stamp Tool

Step 4. When the mouse moves to the screen, its shape will become a hollow circle, and the area enclosed is the area to be copied. As shown in Figure 2-42, hold down the Alt key, click the left mouse key where you want to copy, and select the content to be covered, as shown in Figure 2-43.

Figure 2-42　Mouse Look for Clone Stamp

Figure 2-43　Mouse Look for Sampling

Step 5. Erase on the area to be covered, as shown in Figure 2-44, the image at the sampling point can be copied, as shown in Figure 2-45. After covering, the copied content can be seen on the layer thumbnail, as shown in Figure 2-46.

Figure 2-44　Covering Process

Figure 2-45　After Covering

Figure 2-46　Change in Layer Thumbnail

> Tips:
>
> 1. Create a new layer to clone. If the effect is not ideal, just delete the layer and it will not affect the original image.
>
> 2. The reason for selecting all the layer options in the Sample drop-down list is to let the clone stamp tool obtain the image information from the "merged image of all layers", and then store the information generated by the copying operation in the newly created blank layer; if the default current layer option is used, since the current layer is blank, no matter how it is smeared or erased, nothing can be copied.
>
> 3. It is suggested to enlarge the display scale of the image to facilitate the setting of sampling points and daubing operation.
>
> 4. In the process of smearing, the cross is the sample, which is used to replace the image of mouse position.
>
> 5. When covering large objects, the sampling points can be set several times to make the processing result more natural.
>
> 6. The clone stamp is not limited to the same image. It can also copy the partial content of one image to another image. When copying different images, two images can be placed side by side in the Photoshop window to compare the copy position of the source image and the copy result of the target image.

II. Restore Texture

The clone stamp tool works well, but it has a drawback: when the brightness of the source region is slightly different from that of the target area, the repair result will be very unnatural. Repair tool can overcome this shortcoming. It will repair the texture of the target image, and make the brightness of the repaired area close to the surrounding pixels, making the repair effect more natural.

1. Spot Healing Brush Tool

The smudge repair brush tool is quite magical. It can remove the dirty spots, stains or sundries in the photo. As long as you paint at the place where you want to repair, this tool will automatically take the image content around the repaired area as the repair basis, quickly cover the dirty spots or debris, and can also retain the brightness and shade of the repaired area, so that the repair result does not leave traces. The repair process is as follows:

Step 1. Click 【Create a new layer】 in the Layer Panel to create a new blank layer for drawing the patched content.

Step 2. Select 【Spot Healing Brush Tool】 in Toolbox, as shown in Figure 2-47. In the tool property bar, set the brush to an appropriate size, set the Mode as normal, and select approximate matching for type, and check the box for sampling all layers, as shown in Figure 2-48.

Step 3. After the spot healing brush tool is set, click or erase directly on the stain to complete the repair.

Figure 2-47 Spot Healing Brush Tool

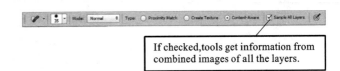

If checked, tools get information from combined images of all the layers.

Figure 2-48 Sample of Property Setting on Spot Healing Brush Tool

Tips:

1. If the repair result of a small defect is not satisfactory, users can erase the defect again, and Photoshop will repair the defect with the texture of different areas around it.

2. If there is no suitable texture in the surrounding area of a defect for sampling, users can select the texture item in the tool property bar, and let Photoshop create a suitable texture to fill in the defect area according to all the contents in the defect area.

3. Smudge repair brush tool is suitable for removing obvious stains or impurities, such as acne, spots on the skin or small scratches on photos. However, since this tool will automatically grab the surrounding image to fill the patched area, if the area to be repaired is complex, it is not suitable to use this tool to repair.

2. Healing Brush Tools

The healing tool is used to repair large areas. The tool is similar to a clone stamp, but it keeps the shading of the patched area, making the repair more natural. The repair process is as follows:

Step 1. Click the 【Create a new layer】 button in the Layer Panel to create a new blank layer for drawing the patched content.

Step 2. In the Toolbox, press and hold 【Spot Healing Brush Tool】, and select 【Healing Brush Tool】 from the opening tool group list, as shown in Figure 2-49. In the tool property bar, set the Brush to the appropriate Sizes, set the Mode to Normal, and set the Source to Sampled. Check the Aligned check box, and select All Layers in the Sample drop-down list, as shown in Figure 2-50.

Step 3. Press the Alt key near the content to be covered and click the left mouse button to set the sampling point, as shown in Figure 2-51.

Step 4. Erase on the content to be covered, as shown in Figure 2-52.

Figure 2-49 Healing Brush Tool

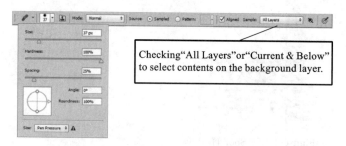

Figure 2-50 Sample of Property Setting on Healing Brush Tool

Figure 2-51 Sampling Process

Figure 2-52 Healing Process

Tips:

1. When the area to be repaired is large, sampling points can be set up several times near the content to be repaired to make the repair more natural.

2. In the process of erasing, sampling points can be set several times to erase the same place to achieve the desired effect.

3. The clone stamp tool can avoid the influence of the surrounding area on the repair results. During the repair process, the clone stamp tool and repair brush tool can be used alternately to make the repair more natural.

4. When dealing with details, enlarge the display scale and reduce the brush to achieve the ratio place to achieve the desired effect.

3. Patch Tool

Patch tool is used to patch the blocked areas. It is suitable for the repair of large and less detailed areas. The repair process is as follows:

Step 1. In the toolbox, press and hold 【Spot Healing Brush Tool】, and select 【Patch tool】 from the open tool group list, as shown in Figure 2-53. The settings of tool property bar maintain the default value, as shown in Figure 2-54.

Step 2. Circle the contaminated area on the image with the repair tool, as shown in Figure 2-55. Drag it to the clean area, as shown in Figure 2-56, and the circle area can be automatically repaired with the clean area, as shown in Figure 2-57.

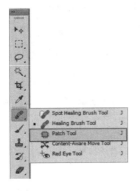

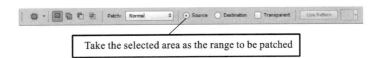

Figure 2-53　Patch Tool

Figure 2-54　Property Bar of Patch Tool

Figure 2-55　Start Circling　　　Figure 2-56　End Circling　　　Figure 2-57　Dragging Process

Tips:

1. When using the Patch Tool, users can also set the patch item as the target in the property bar, then select the target area and drag it to the area to be patched.

2. The Patch Tool only works on the current layer, so it is not allowed to create a new blank layer for repair. If the target area is taken from multiple layers, users can first press the shortcut keys Ctrl+Alt+Shift+e to perform the overlay layer operation. The function of the overlay layer is to merge all the currently visible layers into a new layer, while all the original layers remain unchanged. It can be repaired on the new layer by using the Patch Tool.

Ⅲ. Beautify the Details

After the large-scale beautification of the image, it is often necessary to further beautify the details. The detail beautification of the image includes many aspects, such as adjusting the partial brightness, partial clarity, partial color, and so on.

1. Color Replacement Tool

The color replacement tool replaces one color on the image with another by erasing it. Its essence is to adjust the hue of the pixels in the erasing area of the tool, so the color to be replaced cannot be the colorless phase color such as black, white, gray, etc. The specific operation steps are as follows:

Step 1. Select the area to be replaced and create a selection.

Step 2. Find the 【Brush Tool】 in the 【Toolbox】, press and hold the 【Brush Tool】 button to open the tool group list, and select【Color Replacement Tool】, as shown in Figure 2-58. Set the brush Size, Tolerance and other tool properties in the tool property bar, as shown in Figure 2-59.

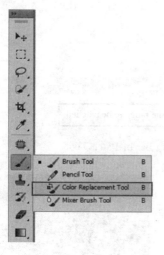

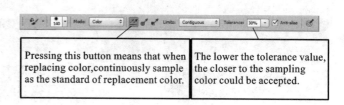

Figure 2-58　Color Replacement Tool

Figure 2-59　Property Bar of Color Replacement Tool

Step 3. Click the color block at the bottom of Toolbox, as shown in Figure 2-60, open the dialog box of Color Picker(Foreground Color), and set the foreground color to the color you want to replace, as shown in Figure 2-61.

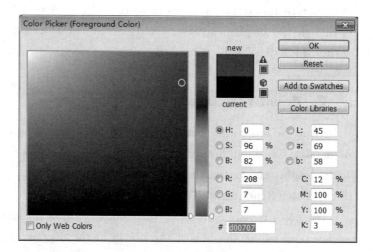

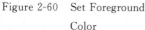

Figure 2-60　Set Foreground Color

Figure 2-61　Color Picker Dialogue Box

Step 4. Erase the selected area with the color replacement tool, as shown in Figure 2-62, and complete the color replacement, as shown in Figure 2-63.

Figure 2-62　Color Replacing Process

Figure 2-63　After Color Replacing

2. Enhance Brightness and Clarity

The partial brightness and definition of the image can be adjusted by using the Dodge Tool and Sharpen Tool respectively. Different from using the menu to adjust, using tools to adjust is to adjust the partial image, that is to say, only adjust the area erased by the tool, and the unprocessed area will not change. The specific operation steps are as follows:

Step 1. Select the 【Dodge Tool】 in the Toolbox, as shown in Figure 2-64. Set the tool property in the tool property bar, as shown in Figure 2-65, and then erase the reflective area on the image to improve its brightness, as shown in Figure 2-66.

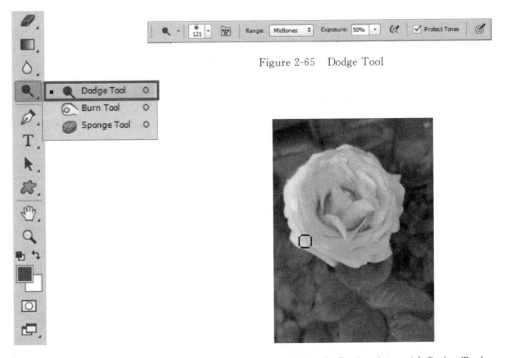

Figure 2-65　Dodge Tool

Figure 2-64　Property Bar of the Dodge Tool　　Figure 2-66　Erase Reflective Area with Dodge Tool

Step 2. Long press and hold【Blur Tool】in Toolbox and select【Sharpen Tool】from the pop-up tool group options, as shown in Figure 2-67. Set the tool properties in the tool property bar, as shown in Figure 2-68, and then erase the object outline on the image to improve the clarity of the drawing surface, as shown in Figure 2-69.

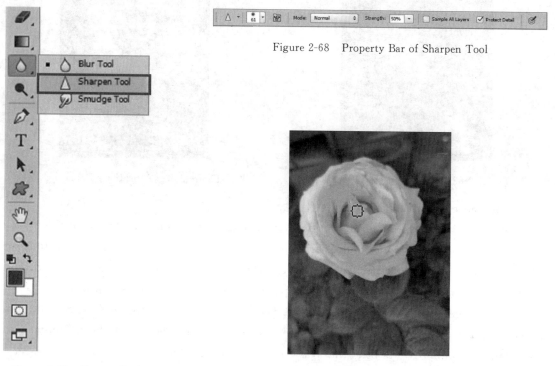

Figure 2-68   Property Bar of Sharpen Tool

Figure 2-67   SharpenTool   Figure 2-69   Erase the Outline with Sharpen Tool

Section 4   Establishment of Selected Area

When editing an image, we often need to process a certain part of the image. When designing an image work, we will also need a part of the image to form a new picture, that is, to pick out the specific content in the image. Selection tools are mainly applied for processing partial image. Photoshop provides a variety of selection tools, so that we can quickly and accurately select images in various situations.

Ⅰ. Rectangular Marquee Tool Group

The rectangular marquee tool group includes rectangle marquee tool, elliptical marquee tool, single row marquee tool and single column marquee tool. This group of selection tools is mainly used to select regions with regular shapes.

In the Toolbox, press and hold the【Rectangular Marquee Tool】button to open the rectangular box selection tool group list, as shown in Figure 2-70. With the rectangular box selection tool and the elliptical box selection tool, users can select the contents of the screen by dragging and drawing the rectangular or elliptical selection box; the single line box selection tool and the single column box selection tool can be used to select the row or column where the click position is located. After selecting a tool, take the rectangular marquee tool as an example to further set the tool properties in the tool property bar, as shown in Figure 2-71.

①The 【New selection】 button is used to create a new area;

②The 【Add to selection】 button is used to add the newly drawn selected area to the existing selected area;

③The 【Subtract from selection】 button is used to subtract the newly drawn selected area from the existing selected area;

④The 【Intersect with selection】 button is used to select the overlapped part between the existing selected area and the newly drawn selected area;

⑤The Style option is used to constrain the shape of the selected area to be drawn, including three drop-down options: the Normal option is to create the selection according to the length and width of the drag; the Fixed Ratio option is to create the selection area according to the setting length-width ratio; the Fixed Size option is to create the selected area according to the setting size, without dragging, just click on the image to form the district with fixed size starting from single-click point as the upper left corner; press Alt and click at the same time to form a district centered on the single-click point.

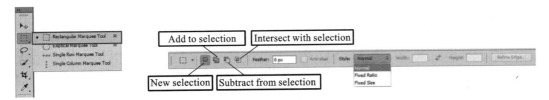

Figure 2-70 Rectangular Marquee Tool Group

Figure 2-71 Property Bar of Rectangular Marquee Tool

Take the rectangular box selection tool as an example, and its usage is as follows:

①Directly drag on the image to form a rectangular selection with the mouse starting point as the upper left corner and the ending point as the lower right corner, as shown in Figure 2-72. The arrow marks the growth direction of the rectangular selection box.

②Hold down Shift and drag on the image to form a square selection with the mouse starting point as the upper left corner and the ending point as the lower right corner, as shown in Figure 2-73. The arrow marks the growth direction of the square selection box.

③Press and hold Alt and drag on the image to form a rectangular selection with the mouse starting point as the center and expanding outward, as shown in Figure 2-74. The arrow marks the growth direction of the rectangular selection box.

④Press and hold Shift + Alt and drag on the image to form a square selection with the mouse starting point as the center and expanding outward, as shown in Figure 2-75. The arrow marks the growth direction of the square selection box.

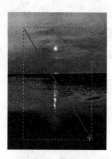 

Figure 2-72 Direct Dragging   Figure 2-73 Hold Shift While Dragging

Figure 2-74 Hold Alt While Dragging   Figure 2-75 Hold Shift+Alt While Dragging

Tips:

1. After selecting the district, drag the mouse in the selected area to move the area.

2. In the process of dragging, don't release the left mouse button, press the space bar to move the selection area. After releasing the space bar, the size of the selection area can be changed.

3. The above function keys only work when there is no selected area in the image. Otherwise, press Shift to add the selected area, and Alt to subtract from the selected area.

The ellipse selection tool is operated with the same method as that of the rectangular box selection tool.

II. Lasso Tool Group

The lasso tool group includes lasso tool, polygonal lasso tool and magnetic lasso tool. This group of tools is mainly applied for selecting the range.

In Toolbox, press and hold the 【Lasso Tool】 button to open the lasso tool group list, as shown in Figure 2-76. The main functions of the three tools in the lasso tool group are mainly 【New selection】, 【Add to selection】, 【Subtract from selection】 and 【Intersection with selection】. Their usage is the same as that in the rectangular selection tool group. The specific operation of each tool in the lasso tool group is as follows:

1. Lasso Tool

【Lasso Tool】 is a tool used to draw the selected area manually. Drag it directly on the image. When release the left mouse button, the tool will automatically connect the starting and ending points to form a closed selection area, as shown in Figure 2-77.

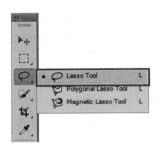

Figure 2-76 Lasso Tool Group

Figure 2-77 Circle with Lasso Tool

2. Polygonal Lasso Tool

The Polygonal Lasso tool is used to select objects with straight lines by clicking. The specific operation steps are as follows:

Step 1. Click anywhere on the polygon to determine the starting point of the selected area, as shown in Figure 2-78;

Step 2. Click at the turning point as the vertex of the polygon, as shown in Figure 2-79;

Step 3. Go back to the starting point and click to form a closed area, or double-click directly to connect the double-click point and the starting point with a line, as shown in Figure 2-80.

Figure 2-78 Starting Point on the Selected Area

Figure 2-79 Turning Point on the Selected Area

Figure 2-80 Ending Point on the Selected Area

Tips:

1. Press Delete during selection to delete the previous vertex.

2. Hold down Shift during selection, and select the next vertex with an integral multiple angle of 45 degrees.

3. Magnetic Lasso Tool

【Magnetic Lasso Tool】 selects the image according to the comparison of adjacent pixels of the image, and the edge of the image will be automatically pasted when selecting, which is suitable for selecting the image with obvious boundary difference in color or brightness. The tool property bar of the magnetic lasso tool is shown in Figure 2-81.

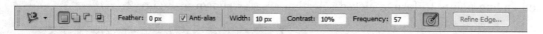

Figure 2-81 Property Bar of Magnetic Lasso Tool

①The meanings of the 【New selection】, 【Add to selection】, 【Subtract from selection】 and 【Intersection with selection】 are consistent with the button functions in the property bar of Rectangular Marquee Tool.

②Width: the magnetic lasso tool takes the mouse pointer as the center to detect the image within a certain distance, so as to find out the edge of the object. Changing the width value can change the detection range. When using this tool, press the Caps Lock key to display the detected area.

③Contrast: set the contrast degree of the object edge as the detection standard. When the edge of the object is clear, the value should be set higher; on the contrary, if the edge of the object is difficult to judge, the value should be set lower.

④Frequency: set the density of nodes during circle selection. The higher the value, the denser the nodes are, and the more accurate the object edge can be selected.

Operational method of the magnetic lasso tool: click at the starting point, and then move the mouse along the edge, as shown in Figure 2-82. Click the mouse at the place with obvious turning point, and manually add nodes.

> Tips:
> During the selection process, press the Delete key to delete the last black control point.

Ⅲ. Quick Selection Tool Group

The quick selection tool group includes quick selection tool and magic wand tool. The tool group mainly establishes selection area based on the screen color.

In the Toolbox, press and hold the 【Quick Selection Tool】 button to open the quick select tool group list, as shown in Figure 2-83. The specific operation of each tool in quick selection tool group is as follows.

1. Quick Selection Tool

【Quick Selection Tool】 allows us to directly erase on the image and create a selection area of similar color ranges, as shown in Figure 2-84. Its property bar is shown in Figure 2-85. Check the 【Auto-Enhance】 option to make the selected area detection more accurate.

2. Magic Wand Tool

With 【Magic Wand Tool】, the selection is realized by clicking. It is used to select pixels whose color is similar to that of the pixel to be clicked. It is very effective when the appearance of the object to be selected is complex. The Magic Wand Tool property bar is shown in Figure 2-86.

①The meanings of 【New selection】, 【Add to selection】, 【Subtract from selection】 and

Figure 2-82 Establish Selected Area with Magnetic Lasso Tool

Figure 2-83 Quick Selection Tool Group

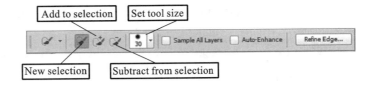

Figure 2-84 Establish Selected Area with Quick Selection Tool

Figure 2-85 Property Bar of Quick Selection Tool

Figure 2-86 Property Bar of Magic Wand Tool

【Intersection with selection】 are consistent with the button functions in that of Rectangular Marquee tTool;

②Tolerance: the value between 0～255 can be input, and the value must be adjusted according to the color difference in the selected image, as shown in Figure 2-87 and Figure 2-88;

③Contiguous: given multiple discontinuous regions in the image are of similar colors, if you cancel this option, all pixels with similar colors at the click point can be selected at one time, as shown in Figure 2-89; if this option is checked, only pixels in the same color region as the click location will be selected, as shown in Figure 2-90;

④Sample All Layers: check this option to select all visible layers at the same time.

Figure 2-87　Establish Selected Area with Tolerance as 32

Figure 2-88　Establish Selected Area with Tolerance as 42

Figure 2-89　Uncheck Contiguous Option

Figure 2-90　Check Contiguous Option

Otherwise, only the currently selected layer will be selected.

Section 5　Multiple Layer Blending

When we need to make some pictures by ourselves, such as posters, PPT background, etc., it is necessary to integrate the collected materials together and display them reasonably on the screen, which requires to control the display relationship between layers, so as to achieve the purpose of multiple layer fusion.

This section introduces the basic methods of making multimedia works by taking the slide background picture shown in Figure 2-91 as an example.

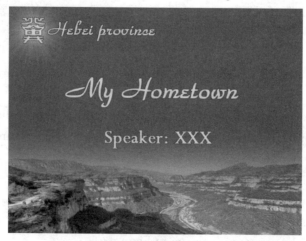
Figure 2-91　Sample Picture with Multiple Layer Blending

Ⅰ. Draw Shapes

1. Create a new file and fill in the background color

Step 1. Click the 【File】 menu and select the 【New】 menu item, as shown in Figure 2-92. In the open 【New】 window, set the parameters of the new document according to the parameters shown in Figure 2-93.

Figure 2-92   New Menu

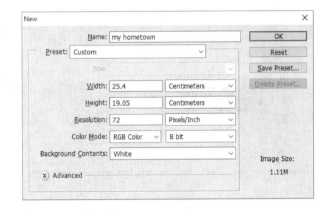

Figure 2-93   New Window

Step 2. Click the 【Set Foreground Color Tool】 at the bottom of the Toolbox, as shown in Figure 2-94, open the Color Picker (Foreground Color) window; in the dialog box, set the color to red(255,0,0), as shown in Figure 2-95.

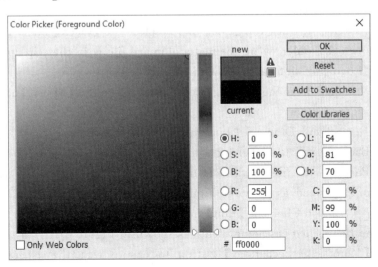

Figure 2-94   Set Foreground
Color Tool

Figure 2-95   Color Picker Window

Step 3. Find 【Gradient Tool】 in Toolbox, press gradient tool, and select 【Paint Bucket Tool】 from the open tool group list, as shown in Figure 2-96; click on the canvas with paint bucket tool to paint the canvas color red, as shown in Figure 2-97.

Figure 2-96   Paint Bucket Tool   Figure 2-97   Paint Background

Tips:
　　1. Pay attention to the unit when setting the size of new file.
　　2. The color mode of the new file should be set to RGB mode.
　　3. The square color in the tools bar is the current foreground color, not necessarily black.

2. Insert Logo

Step 1. Click the 【File】 menu, select the 【Open】 menu item, as shown in Figure 2-98, open the folder where the material image file is in the pop-up 【Open】 window, and find the "Logo. png" image, as shown in Figure 2-99.

Step 2. In the opening "Logo" file, select the 【Move Tool】 in 【Toolbox】, as shown in Figure 2-100.

Step 3. Drag the "Logo" image, as shown in Figure 2-101, and stay on the "Cover Background" file label for a moment, as shown in Figure 2-102. After opening the "Cover Background" file, continue to drag downward, and place the "Logo" in the upper left corner of the screen, as shown in Figure 2-103.

Step 4. Click the 【Edit】 menu, select the 【Transform】 item, and select the 【Scale】 item in the opened cascading menu, as shown in Figure 2-104; in the tool property bar, press the 【Maintain aspect ratio】 button to ensure that the aspect ratio of the image remains unchanged. Enter 25% in the text box after the width W to reduce the image to 25%. After setting, click the "√" button on the right side of the attribute column determines the adjustment, as shown in Figure 2-105.

Figure 2-98　Open Menu

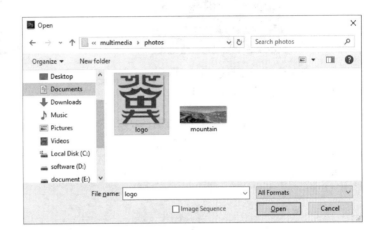

Figure 2-99　Open Window

Figure 2-100　Move Tool

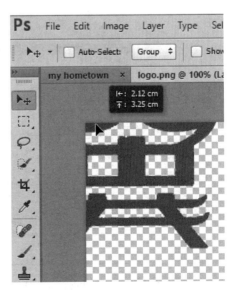

Figure 2-101　Drag "Logo"

Figure 2-102 Stay on the Cover Background

Figure 2-103 Move "Logo" on the Upper-left Corner

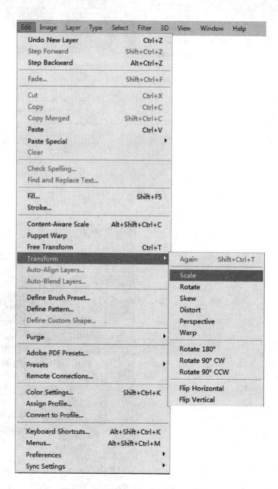

Figure 2-104 Scale Menu Item

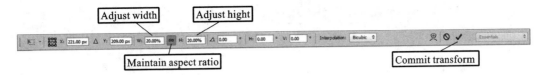

Figure 2-105 Property Bar of Scale Setting

> Tips:
> After selecting the Scale, eight control points will appear around the layer. Dragging the control points can also adjust the size of the object. This adjustment can not determine the scale of the object, but only determine whether the size is appropriate according to the result of dragging.

3. Draw Light

Step 1. Click the 【Create a new layer】 button in the layer panel to create a new blank layer for drawing light. Double click the layer name to name the layer as "Light", as shown in Figure 2-106.

Step 2. Click the foreground color tool at the bottom of Toolbox to open the Color Picker (Foreground color); in the dialog box, set the color to yellow(255,255,0).

Step 3. Find the 【Rectangle Tool】 in the Toolbox, press and hold it to open the tool group list, and click the 【Custom Shape Tool】 in the list, as shown in Figure 2-107; in the tool property bar, choose Pixels in the drop down list box of 【Pick tool mode】, click the lower triangle on the right side of Shape, and click the desired shape in the shape selection list, as shown in Figure 2-108.

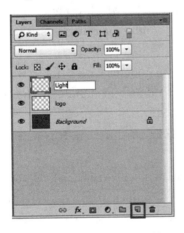
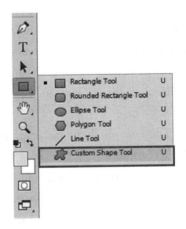

Figure 2-106 Change Layer Name　　　　　Figure 2-107 Custom Shape Tool

Step 4. Drag and drop the appropriate position on the canvas to draw the light pattern of appropriate size, as shown in Figure 2-109.

Step 5. On the layer panel, drag down the "Light" layer to make it under the "Logo" layer, as shown in Figure 2-110; use the 【Move Tool】 to drag the light pattern on the canvas to make it under the "Logo", as shown in Figure 2-111.

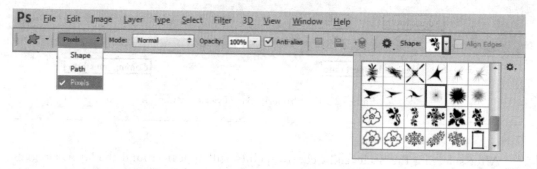

Figure 2-108　Property Bar of Custom Shape Tool

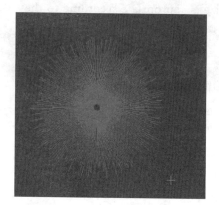

Figure 2-109　Draw Pattern　　　Figure 2-110　Adjust Layer Sequence　　　Figure 2-111　Move Layer

Tips:

1. A new blank layer must be created to draw the light, which cannot be drawn on the university emblem layer. Otherwise, it will be very troublesome to adjust the relative position of the university emblem and the light.

2. In the property bar of Custom Shape Tool, the function of Path is to draw the path of the selected shape, and the function of Pixels is to draw the figure of the selected shape.

3. The custom shape library can be downloaded from the Internet. It is usually a file with ". csh" extension.

4. Loading method of custom shape library: in the tool property bar of Custom Shape Tool, click the lower triangle on the right side of shape to open the shape selection list. Click the setting icon ✿. at the top right of the list to open the cascading menu, select the Load Shapes, open the load window, find the custom shape file in the computer, and click Load to load the shape library into the shape selection list.

Ⅱ. Input Text

There are two ways to input text in Photoshop: anchor text and paragraph text. The input method of anchor text is to click the place where you want to add text in the image, and then type the text after the insertion point appears, and users can decide when to wrap the text. If it is expected to fix the text in a certain range, users can use paragraph text input mode, that is, drag out a text box, and then enter the text, so that the text will be automatically arranged in the text box, line feed. The operation steps are as follows:

Step 1. Select 【Horizontal Type Tool】 in Toolbox, as shown in Figure 2-112, and set the font, font size, color and other tool properties in the tool property bar, as shown in Figure 2-113.

Step 2. Click to create the anchor text input box where the text needs to be displayed, or drag the appropriate rectangular area to create the paragraph text input box.

Step 3. Input the required text in the input box, and then click "√" on the options bar to confirm the input. At this time, a text layer with the input text as the layer name will be automatically created on the layer panel.

Step 4. With the same method, input speaker and Province name, as shown in Figure 2-114.

Figure 2-112　Horizontal Type Tool

Ⅲ. Layer Blending

Figure 2-113　Property Bar of Horizontal Type Tool

Figure 2-114　Background with Text

1. Layer blending mode

The layer blending mode refers to that when the image is superimposed, the pixels of the upper layer and the lower layer are mixed to obtain another image effect. Therefore, the layer blending mode can only work between two layers of a image. In the case of mixing the

"Mountain" layer and the background layer, the operation steps are as follows:

Step 1. Click the 【File】 menu, select the 【Open】 menu item, open the "Mountain" image, and drag the "Mountain" image to the bottom of the "Cover Background" canvas, as shown in Figure 2-115.

Step 2. On the layer panel, adjust the "Mountain" layer to the top of the "Cover Background" layer, as shown in Figure 2-116. Then, in the Layers drop-down list at the top left of the panel, set the layer blending mode to Luminosity, as shown in Figure 2-117, to realize the mixing of the mountain layer and the background layer, as shown in Figure 2-118.

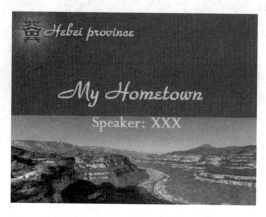

Figure 2-115 Background with "Mountain" Layer

Figure 2-116 Adjust Layer Sequence

Figure 2-117 List of Layer Blending Mode

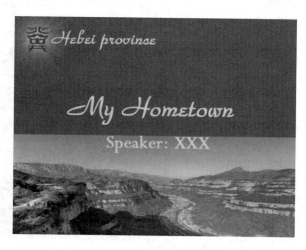

Figure 2-118 After Layer-blending

Tips:

1. The functions of the options in the layer blending mode list are as follows:

(1) "Normal" mode: as the default mode of Photoshop, it does not produce any special effects when used.

(2) "Dissolve" mode: when this option is selected, the image screen will produce dissolution and granular effect. The smaller the opacity value on the right side of the layer panel, the more obvious the dissolution effect is.

(3) "Darken" mode: if this option is selected, the software will take the dark color of two colors as the final color when drawing the image, the color that is lighter than the background color will be replaced, and the color that is darker than the background color will remain unchanged.

(4) "Multiply" mode: select this option to generate a color that is darker than the background color and the drawing color, and can be used to make shadow effects.

(5) "Color Burn" mode: select this option to deepen the color of the image and reduce the brightness of the image.

(6) "Linear Burn" mode: select this option to darken the background color by decreasing the brightness, thus reflecting the color of the painting. When mixed with white, it does not change.

(7) "Darker Color" mode: when selected, the system selects the smallest channel value from the base and blend colors to create the result color.

(8) "Lighten" mode: this mode only works if the current color is darker than the base color, and the light color will cover the dark color.

(9) "Screen" mode: this option has the reverse function of the positive overlay, which is usually lighter in color. When the base color is mixed with the painted black, and the original color is not affected; if it is mixed with the painted white will be generated; if it is mixed with other colors, a bleaching effect will be obtained.

(10) "Color Dodge" mode: select this option to reflect the color of the drawing by reducing the contrast and making the background color brighter. There is no change when blending with black.

(11) "Linear Dodge" mode: select this option to make the background color brighter by increasing the brightness, so as to reflect the painted color. There is no change when blending with black.

(12) "Lighter Color" mode: select this option to choose the largest channel value from the base and blend colors to create the result color.

(13) "Overlay" mode: select this option to make the pattern or color superimposed on existing pixels while preserving the light and shade contrast of the base color.

(14) "Soft Light" mode: if you select this option, the system will decide whether to lighten or darken according to the shading of the painted color. When the painted color is

darker than 50%, the image is dimmed by increasing the contrast.

(15) "Hard Light" mode: if you select this option, the system will decide whether to perform positive lamination or filtering according to the mixed color. However, when the painted color is lighter than 50% gray, the background image is lighter; when it is darker than 50% gray, the background image becomes dark.

(16) "Vivid Light" mode: select this option to darken or lighten colors by increasing or decreasing contrast based on the painted color. When the drawn color is darker than 50% gray, the image is dimmed by increasing the contrast.

(17) "Linear light" mode: when this option is selected, the system also deepens or lightens the colors by increasing or decreasing the brightness according to the painted color. When the painted color is lighter than 50% gray, the image is lightened by increasing brightness, and when it is darker than 50% gray, the image is dimmed by decreasing brightness.

(18) "Pin light" mode: select this option to replace the color based on the painted color. When the painted color is lighter than 50% gray, the former is replaced, but the pixels brighter than the painted color are not replaced; when the painted color is darker than 50% gray, the pixels brighter than the painted color are replaced, but the pixels darker than the painted color are not replaced.

(19) "Hard Mix" mode: select this option to add the red, green, and blue channel values of the blend color to the RGB value of the base color. If the sum of the channel calculation is greater than or equal to 255, the value is 255; if it is less than 255, the value is 0.

(20) "Difference" mode: if you select this option, the system will subtract the dark pixel value from the brighter pixel value, and the difference will be the final pixel value. When mixed with white, the background color will be reversed, while mixed with black, nothing will be changed.

(21) "Exclusion" mode: select this option to bring out a similar effect to the "normal" option, but with a smaller color contrast than the difference mode, so the color is softer.

(22) "Hue" mode: select this option, the system will adopt the brightness and saturation of the lower image, and the hue of the upper image to create the final color.

(23) "Saturation" mode: select this option, the system will use the brightness, hue of the lower image, and the saturation of the upper image to create the final color.

(24) "Color" mode: select this option, the system will adopt the brightness of the lower image and the hue and saturation of the upper image to create the final color.

(25) "Luminosity" mode: select this option, the system will adopt the hue, saturation of the lower image and the brightness of the upper image to create the final color. This option achieves the opposite effect as the "Color" option.

> 2. In the above 25 options of layer blending mode, RGB color model is used in 1 $\sim$ 21 and HSB color model is used in 22 $\sim$ 25.

2. Layer mask

(1) Advantages

The biggest advantage of layer masking is non-destructive editing. Using layer mask allows users to change the image without covering the original image data, and keep the original image data available for recovery at any time. Since non-destructive editing does not remove data from the image, it does not degrade image quality when editing.

(2) Function

Layer mask is the most commonly used mask in image processing. It is mainly used to show or hide part of the content of the layer, and keep the original image from being damaged by editing.

The white area in the layer mask can cover the content of the following layer and only display the image in the current layer; the black area can cover the image in the current layer and display the content in the lower layer; the gray area in the mask can make the image in the current layer present different levels of transparency according to its gray value.

The so-called layer mask can be understood as a piece of cloth that can make the object transparent. When the cloth is coated with black, the object becomes transparent; when the cloth is coated with white, the object is completely displayed; when the cloth is coated with gray, the object shows translucent effect.

In fact, the layer mask is a gray image. When it is combined with the layer image, different grays will have different degrees of masking effect.

- The masking rate of black is 100%, so if the layer image corresponds to the black part, it will become completely transparent and reveal the content of the lower layer image.
- The masking rate of white is 0%, so if the layer image corresponds to the white part, it will remain unchanged, that is, it is completely opaque.
- The masking rates of gray with different degrees are also different. The closer to the black, the higher the masking rate is, the more transparent the layer image is. The closer to the white, the lower the masking rate is, the more opaque the layer image is. Therefore, if the layer image corresponds to the gray part, it will have different degrees of translucent effect.

(3) Operation Steps

Step 1. On the layer panel, select the layer to be added with mask, and click 【Add vector mask】 at the bottom of the panel to add a layer mask for the current layer. At this time, a mask thumbnail will be associated to the right of the layer thumbnail, as shown in Figure 2-119.

Step 2. In the Toolbox, find the 【Gradient Tool】, press and hold the tool to open the tool group list, and select it, as shown in Figure 2-120. In the tool property bar, click the lower triangle on the right side of the gradient ribbon to open the gradient picker, select Black, White gradient, and select 【Linear Gradient】 for the gradient style on the right, as shown in

Figure 2-121.

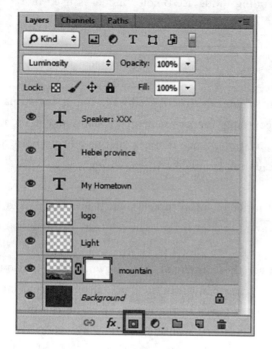

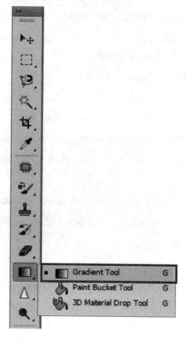

Figure 2-119 Add Vector Mask

Figure 2-120 Gradient Tool

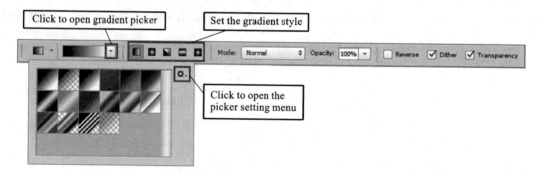

Figure 2-121 Property Bar of Gradient Tool

Step 3. Drag the "Mountain" position on the canvas from top to bottom to draw a line, as shown in Figure 2-122. During the drawing process, press the Shift key at the same time to ensure that the drawn line is perpendicular to the horizontal line. The effect picture is shown in Figure 2-123, and the layer panel status is shown in Figure 2-124.

> Tips:
> 1. Create a selection area, select the required screen content, and add a layer mask to "hide" the unnecessary screen content to achieve matting.
> 2. Various Photoshop tools can be used to edit pixels, such as brush tools, gradient tools, various filters, etc., to directly edit layer masks and create various synthesized effects.

Figure 2-122　Gradient Fill

Figure 2-123　After Adding Vector Mask

Figure 2-124　Layer Panel after Adding Vector Mask

Ⅳ. Layer Style

Layer style can quickly change the appearance of layer content, and make image effects such as projection, external illumination, overlay, stroke, etc. The specific operation methods are as follows:

Step 1. In the layer panel, select the layer to be applied, such as the layer of "My Hometown", and then double-click the layer to open the Layer Style dialog box, as shown in Figure 2-125.

Step 2. In the Styles on the left side of the Layer Style dialog box, click the Stroke option, and the right side displays parameter setting interface of the stroke effect. Drag the slider on the right side of Size or directly enter a value in the text box to set the edge to 2 pixels wide. Click the color block behind Color to adjust the edge color to black(0,0,0), as shown in Figure 2-126.

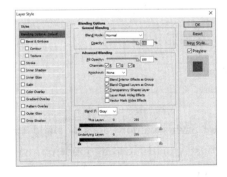 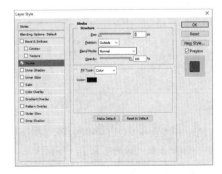

Figure 2-125　Layer Style Dialog Box

Figure 2-126　Stroke Setting

Step 3. With the same method, set the "stroke" effect for "Speaker: XXX" and "Hebei province".

Step 4. Double click the "logo" layer to open the Layer Style dialog box, select the Color Overlay option on the left, click the rectangular color block in the parameter setting interface on the right, as shown in Figure 2-127, open the Select overlay color window, and set the overlay color to golden yellow(255,255 0),as shown in Figure 2-128;select the Stroke option to set the stroke effect of this layer consistent with that of other layers.

Figure 2-127 Set Color Overlay

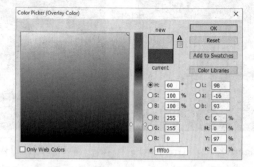

Figure 2-128 Select Overlay Color Window

Step 5:After setting the layer style,the background image of the slide is completed. Finally,click the 【File】 menu,select the 【Save】 or 【Save as】 menu item,open the window as shown in Figure 2-129,select the file format of the output image,input the file name,select the file storage path,and click the 【Save】 button to save the image file.

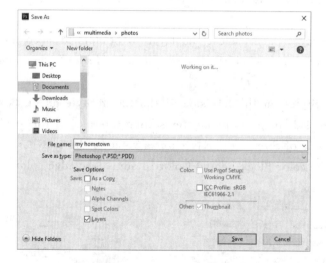

Figure 2-129 Save Window

Tips:

1. The top left[Blending Options]in the[Layer Style]window does not add layer style effects to layers. It is mainly used to adjustthe opacity and blending mode of layers, and control the blending mode of pixels between layers.

(1) The options area of[General Blending]is mainly used to set the opacity and blending mode of layers.

(2) [Fill Opacity]in the[Advanced Blending]options area is used to adjust the filling opacity degree of a layer. Unlike opacity in the[General Blending]option,filling opacity affects pixels in the layer or shapes drawn on the layer,but does not affect the opacity of

the added layer style.

(3) [Channels] is used to select the channels participating in the mixing. By default, all channels are selected. Users can also specify channels.

(4) [Knockout] is used to select a hollowing method. After setting the knockout method, users can see the contents of the layer below through a certain layer.

(5) When the [Blend Interior Effects as Group] check box is selected, when the "Inner Glow", "Stain", "Color Overlay", "Gradient Overlay", and "Pattern Overlay" styles are added to the layer, the effects are not displayed if [General Blending] is set.

(6) When the [Blend Clipped Layers as Group] check box is selected, knockout is only valid for clip layers; this option also controls the blending mode of the clip mask. When this option is selected, the blending mode of the base layer will be applied to all layers in the clip mask.

(7) Select the [Transparency Shapes Layer] check box to restrict layer and knockout effects to opaque areas of the layer.

(8) Select the [Layer Mask Hides Effects] check box to limit the layer effect to the area defined by the layer mask, and the effect will not be displayed in the layer mask.

(9) Select the [Vector Mask Hides Effects] check box to limit the layer effect to the area defined by the vector mask, and the effect will not be displayed in the vector mask.

(10) The slider in the blend color band under [Blend If] item controls the pixels that displayed by the current layer and the layers below in the final image. The range of partially mixed pixels can also be defined to produce a smooth transition between the mixed and non blended regions. In the drop-down list of the [Blend If], users can select the channel to blend, and selecting [Gray] means to blend with all color channels. [This Layer] is the current layer. Drag the [This Layer] slider to control the range of pixels in the current layer to be mixed and appear in the final image. It ranges from 0 (black) to 255 (white). Drag the black slider to set the lowest value of the range, and drag the white slider to set the highest value of the range. [Underlying Layer] is the layer below the current layer. Drag the slider in the [Underlying Layer] to set the range of pixels in the underlying layer to be mixed and appear in the final image.

2. The functions of [Styles] option on the left of the [Layer Style] window are as follows:

(1) The [Drop Shadow] effect can add a projection effect to the layer content to make it have a three-dimensional effect.

(2) The [Inner Shadow] effect can increase the projection effect inside the edge of the image in the layer, so that the image has a three-dimensional and concave visual sense.

(3) The [Outer Glow] effect creates a glow out along the edge of the layer content.

(4) The [Inner Glow] effect creates a glow inward along the edge of the layer content.

(5) The [Bevel and Emboss] effect can add various combinations of highlights and shadows to the layer, so that the layer content presents a three-dimensional relief effect.

(6) The [Stain] applies smooth glossy interior shadows, usually used to create a glossy appearance on metal surfaces.

(7) The [Color Overlay] effect can overlay a specified color on a layer, and you can control the overlay effect by setting the color blending mode and opacity.

(8) The [Gradient Overlay] effect overlays a specified gradient color on a layer.

(9) The [Pattern Overlay] effect overlays a specified pattern on a layer and scales the pattern to set the opacity and blending mode of the pattern.

(10) The [Stroke] effect can use colors, gradients, or patterns to outline objects, and is especially useful for hard edged shapes such as text.

3. When outputting image files, JPEG image file format is generally selected. When the background of image works is transparent, PNG image file format should be used. It is recommended to store an image file in PSD format in addition to the ordinary image file. Because the Photoshop source file stored in this format retains editing information such as layers and channels, it is convenient to conduct secondary image processing in the future editing.

Chapter 3 Operations of Digital Audio

The sound exists everywhere in our life, for instance, the noise of vehicles, the voice of people, the sound of nature, and all kinds of sound. So is the case in multimedia. The appearance of sound makes the multimedia more colorful. The sound is produced by vibration and transmitted through the air. Multimedia audio is the transformation of the air vibration of sound into a continuously changing electrical signal, and then the signal is sampled, quantified, encoded, and recorded in the form of a file. People record the sound by using digital tape recorders, laser recorders and optical MD recorders as carriers. Now, more people use the computer's hard disk as a carrier to record sound, which is the innovative way of recording sound. Today, sound signals processes by multimedia technology can range from 20 to 20,000Hz.

Section 1 Fundamental Knowledge of Audio

I. Fundamental Knowledge of Sound

Tone, timbre, and loudness are the basic characteristics of sound. In addition to making its own pure sound, the sound source is accompanied by different frequencies of overtones that show the different properties of the sound source object. The sound processed by the multimedia system is the audio within the range of human ears, which is stored as a file in different formats.

1. The Property of Sound

Sound is a physical phenomenon caused by the vibration of an object. Vibrations cause the air around an object to move and then sound waves are formed, which are transmitted to the human ear by the medium of air, then the sound is heard. Therefore, sound is a kind of wave in physical nature. Using a physical method of analysis, the physical quantities of sound features are amplitude, period, and frequency. As frequencies and periods are mutual reciprocals, sound is generally described only by two parameters: amplitude and frequency.

It should be noted that a real-world sound is not composed of waves of a certain frequency or several frequencies, but is composed of many sine waves of different frequencies and amplitudes. Therefore, there will be the lowest and highest frequencies in a sound. Popularly speaking, the frequency indicates the pitch of the sound while the amplitude reflects the volume of the sound. The more high-frequency components the sound contains, the higher(or sharper) the tone, vice versa; the higher the amplitude of vibration is, the louder the sound, and vice versa.

2. The Categorization of Sound

There are several standards for the categorization of sound. Mainly it can be divided into three categories according to different objectives.

(1) Categorization by the standard of frequency

①Infrasound: 0~20 Hz

②Audio: 20Hz~20k Hz

③Ultrasound: 20k Hz~1G Hz

④Hyper-sound: 1G Hz~1T Hz

The purpose of categorization by the standard of frequency is to distinguish the audio heard by human ears and the non-audio sound beyond the hearing ability of humans.

(2) Categorization by the standard of source sound

①Speech sound: it refers to the sound with which humans express their thoughts and emotion.

②Sound of musical instrument: it refers to the sound of musical instrument.

③Sound and noise: it refers to the other sound except for the speech sound and sound of musical instrument, the sound or noise from the nature or certain objects such as sound of wind and rain, thunderstorm.

The distinction of sound from different sources is tofacilitate the digitization of different sampling frequencies for different types of sounds, and different recognition, synthesis and coding methods are used according to the methods and characteristics of their production.

(3) Categorization by the standard of storage form

①Simulation sound: store the sound with simulating method, for instance, record the sound with tapes.

②Digital sound: process the simulated sound digitally and use 0 and 1 to express the data flow or the speech sound and music processed by the computer.

3. Three Elements of Sound

Sound in nature is caused by the vibration of an object, and the object that is making sound is called a sound source. The number of times an object vibrates within a second is called frequency, in hertz(Hz). A person's ear can hear sound from 20 Hz to 20,000 Hz.

(1) Tone

Tone is one of the subjective properties of sound. It is determined by the frequency of an object's vibration, and is related to the intensity of sound. As to the pure sound of the same intensity, the tone rises and falls with the rise and fall of frequency, while the pure sound of certain frequency and the tone of low-frequency pure sound decrease with the increase of sound intensity, while the tone of high-frequency pure sound increases with the rise of with the increase of intensity.

The high and low pitch isalso related to the structure of the sound body, which affects the frequency of sound. In general, the tone of low-frequency pure sound below 2000Hz decreases with the increase of loudness, and the tone of high-frequency pure sound above 3000Hz increases with the increase of loudness.

The commonly said three-degree-sound, bass, median pitch and high pitch, refers to the highness of tone, that is the highness of frequency of sound wave. The high-frequency sound is high pitch and the low-frequency sound is low pitch. In the musical notation we know that the high-pitched "1" is eight degrees higher than the mid-pitched"1" and the mid-pitched"1" is eight degrees higher than the bass "1". A sound used as a sounding standard in music is the one with a frequency of 440Hz, that is, the C tone "6" sound. Then the frequency of the mid pitched"6" is 880Hz, the high-pitched "6" is 1760Hz. This means that the octave-degree frequency is twice as high as it is, and it can be seen that the scale in the music score is not an arithmetic progression but a geometric progression. We know that there are five semitones in the seven sounds from "1" to "7", that is, a total scale of 12 ones, each called a "half tone". Obviously, the ratio of the second semitone is 2 of 12 square roots (about 1.059463), from which we can calculate that the mid-pitched "6" ascending semitone is 932Hz, the mid-pitched "7" is 988Hz, etc.

(2) Timbre

Timbre is the feature of sound, and it differs from the different sound made by the different objects. The sound wave shows the difference and feature of different sound. Figure 3-1 is the sound wave of sound made by different musical instruments.

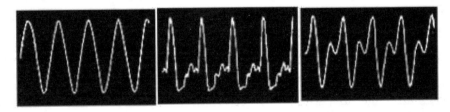

Figure 3-1 Wave of Sound Made by Different Musical Instruments

The difference in timbre lies in the different overtones. In the sound of each instrument, different people and all the sounding objects, in addition to a base sound, there are many different frequencies of the overtone accompanied. It is these overtones that determine the different timbre, so that people can distinguish between different instruments and different sounds made by different people.

(3) Loudness

Loudness is the strength of the sound, also known as volume. It is related to the intensity of sound vibration and the distance of transmission. The greater the intensity of sound source vibration, the greater the loudness, the farther the distance of sound transmission.

The loudness of sound depends on the conditions such as intensity, pitch, timbre and length. If other things are the same, vowels sound louder than consonants; the loudness of vowels is related to the size of the opening of mouth. Vowels with large openings are louder. As to the consonants, voiced sound is louder than the voiceless one while the aspirated sound is loudness than the non-aspirated one.

1) The unit of volume—decibel

The function of the human ears is very special in that human ear can hear a steel needle landing sound as well as bear the sound of thunder. Why is the human ear adaptation range so large? According to the research of the scientists, vibration takes place when the eardrum is affected by the sound and the range of vibration is not in proportion with the intensity of sound but it forms direction ration with the logarithm of sound intensity.. Therefore, the base-10 logarithm of sound intensity is taken, and the "1" value is the unit, which is called "Bell". One-tenth of "Bell" is called "decibels", which is expressed as "db".

The sensitivity of the human ear to sound waves at different frequencies is different. Human ears are most sensitive to sound waves of 3000Hz. As long as the sound of this frequency reaches 10-12 watts/square meters, it can cause hearing in the human ear. The sound intensity level is based on the minimum sound intensity that the human ear can hear, and this sound intensity is specified as zero sound intensity, that is to say, the sound intensity level is zero Bell(also zero dB).

2) The common environment noise value

Generally speaking, a quiet environment we recognize refers to the sound intensity of less than 20 decibels, and the sound intensity that is below 15 decibels is regarded as "dead silence." Whispering is about 30 dB. 40-60 decibels is our normal conversation. Decibels below 60 belong to harmless area, decibels above 60 belong to the noisy range. The sound of 70 decibels is regarded as noisy one, and it begins to damage the hearing nerve. The sound above 90 decibels will impair hearing ability. The car noise is between 80-100 decibels, taking the noise of 90 decibels from a car as an example, 81 decibels sound can still be heard at 100 meters(the above standards will vary from environment to environment, not the absolute value)distance. The chainsaw sound is about 110 decibels and the sound of the jet is about 130 decibels.

II. Digitization of Sound Signal

The digitization of sound signal is the sampling and quantification of the sound signals that fluctuate continuously in time, encoding the quantitative results with some kind of audio encoding algorithm, and the result is the digital form of the sound signal. That is, the sound (analog) is converted at regular intervals into a discrete sequence represented by a limited number of numbers, i. e. digital audio frequency. See Figure 3-2.

Figure 3-2 Digitization of Sound Signal

1. Sampling and Frequency of Sampling

Sampling, also known as taking samples, is the signal that turns a continuous analog signal over time into a limited number of sample values that are intermittently discrete in time, see Figure 3-3. Suppose the sound wave is shown as Figure 3-3(a), and it is the continuous function of time $x(T)$. If sampling is needed, the range value is got from the sound wave according to certain time interval(T), then a sequence $x(nT)$ is got, that is, $x(T), x(2T), x(3T), x(4T), x(5T), x(6T)$, etc. T is called sampling period and $1/T$ is sampling frequency. Sequence $x(nT)$ is the discrete signal of the continuous wave. It is obvious that the discrete signal is the limited number of vibration range sampling value got from the continuous signal $x(T)$.

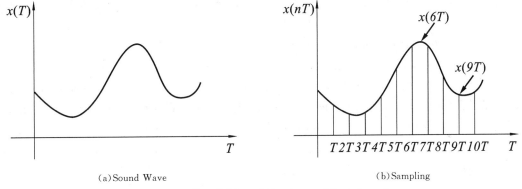

(a) Sound Wave   (b) Sampling

Figure 3-3 Sketch Map of Continuous Wave Sampling

According to the Nyquist sampling theory, the amount of information will not be lost as long as the sampling frequency is equal to or greater than twice of the highest frequency component in the sound signal. That is to say, only when the sampling frequency is twice of or twice more than the maximum frequency of the sound signal, can the sound represented by the digital signal be restored to its original sound (the original continuous analog audio signal), otherwise it will produce varying degrees of distortion. The law of sampling is expressed as:

$$f_{sampling} \geq 2f \text{ or } T_{sampling} \leq T/2$$

In the above equation, f is the highest frequency of sound signal. If the highest frequency of sound signal is f_{max}, then the minimum sampling frequency should be $2 f_{max}$.

A well-known example of Nyquist's sampling theory is the telephone and CD records used in everyday life. The signal frequency of the telephone voice is approximately 3.4kHz. In the digital telephone system, in order to turn the human voice into digital signal, pulse-coded modulation(PCM) method can be used to sample 8000 times per second. CD records store digital information. To achieve the effect of CD sound quality, a sample frequency of 44.1k Hz is required, i. e. a signal with a frequency of up to 22.05k Hz can be captured. In multimedia technology, three audio sampling frequencies, 11.025k Hz, 22.05k Hz and 44.1k Hz, are typically used. Under conditions that allow distortion, the sampling frequency should be selected as low as possible to avoid taking up too much data.

The common audio sampling frequencies and applications are as follows:

①8k Hz, suitable for voice sampling. It can meet the standard of telephone voice quality.

②11.025k Hz, used for voice and the highest frequency of no more than 5k Hz sound sampling. It can reach the standards of telephone voice quality above, but not meet the amplitude broadcast sound quality requirements.

③16k Hz and 22.05k Hz, suitable for the highest frequency of no more than 10k Hz sound sampling. It can reach the amplitude broadcast sound quality requirements.

④37.8k Hz, suitable for the highest frequency of no more than 17.5k Hz sound sampling. It can reach the amplitude broadcast sound quality requirements.

⑤44.1k Hz and 48k Hz, mainly used for music sampling, and it can reach the quality standard of laser disc. For the sound with highest frequency below 20k Hz, sampling frequency of 44.1k Hz is used so as to reduce the storage space of digital sound.

2. Quantization and Quantization Bits

Sampling only solves the digital problem of splitting a waveform into equal parts on a time coordinate(i. e. horizontal axis), but what is the height of the rectangle for each equal part? That is, some digital method is needed to reflect the sound wave voltage value of each instant. The size of this value directly affects the volume. We call the digitized representing of the amplitude of sound waveforms quantization.

The process of quantization is like this: first, divide the sampled signal into a limited set of segments by the amplitude of the entire sound wave, then put the sample values that fall into a segment into one group, and assign them the same quantization values. How to divide the range of sampling signal? Binary method is used in which the vertical axis is divided by the eight bit or sixteen bit. That is, for the sound effect with eight bit recording mode, the vertical axis will be divided into several quantization levels to record the range. For the sound effect with sixteen bit recording mode, the sound range selected in a certain sampling section will be recorded with several different quantization levels.

The sketch map of the sampling and quantization of sound is shown in Figure 3-4.

3. Coding

When the analog signal is sampled and quantified, a series of discrete signals are formed-pulsed digital signals. This pulsed digital signal can be encoded in a certain way to form data used in the computer. Coding is the process of expressing binary data of quantitative results in a certain format. That is, in a certain format, the sampled and quantified discrete data is recorded, and some data for error correction, synchronization, and control is added to the useful data. When the data is playback, it can determine whether the read-out sound data is

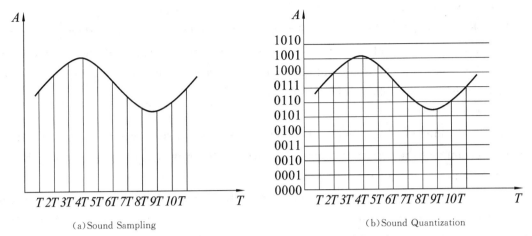

(a) Sound Sampling (b) Sound Quantization

Figure 3-4 Sampling and Quantization of Sound

wrong based on the error correction data recorded. If something is wrong within a certain range, it can be corrected.

There are many forms of coding, and the commonly used coding method is PCM-pulse coding modulation.

Ⅲ. Storage of Audio File

1. The Volume of Audio File

The storage amount of digital audio is determined by the sampling frequency, sampling accuracy, and number of channels for analog sound waves.

The storage amount = sampling frequency × sampling accuracy × number of channels/8 (B/s).

The unit of sampling frequency is Hz and the sampling accuracy is bit. The purpose of dividing the product by 8 is to convert bits into bytes.

For example, the sampling frequency is 44.1k Hz, and the sampling accuracy is 16b, then the amount of data for one second of stereo audio is calculated as follows:

The storage amount = 44100 × 16 × 2/8 = 176.4k B/s

2. Format of Audio File

Digital audio is encoded in digital audio formats, and different digital audio devices generally correspond to different audio file formats. Common audio formats are CD format, WAVE format, MP3 format, MIDI format, WMA format, Real Audio format, etc.

(1) CD format

The audio file with CD format is a kind of audio format with high sound quality that the sampling frequency is 44.1K Hz and the rate is 88K Hz/second. The ".cda" file in CD is only index information instead of real sound one. This information can be converted into other audio formats such as MP3 with *Format Factory*.

The sound channel of CD is approximately lossless as its sound is close to the acoustic. The CD disk can be played in the CD recorders as well as the player of the computer.

(2) WAVE format

WAVE is the kind of audio file format supported by Windows platform and other application. It supports multiple audio digits, sampling frequency and sound channel. The sampling frequency is 44.1K Hz and the rate is 88K Hz/second. The file has ".wav" as suffix, which is widely applied in PC.

(3) MP3 format

MP3 is the kind of audio compression technology which takes advantage of MPEG Audio Layer 3 to compress the music into smaller file with 1:10 or even 1:12 compression ratio and keep the original sound quality.

MP3 is the three layers of standard audiofrequency MPEG-1. According to the complexity and efficiency of encoding, it is divided into three layers: layer I, layer II and layer III. The compressing code of MP3 combines the two calculations of MUSICAM and ASPEC, which greatly improves the compression rate of files, but also ensures the quality of audio.

(4) MIDI format

MIDI originally refers to digitalized instrument interface, which is the name of an interface for signal transmission by difference devices. As the early electronic synthesis technical specifications are not uniform until MIDI 1.0 appeared, all the digital instrument adopt this uniform specification to transmit MIDI information, which forms the combined musical performance system. It is regarded as computer music.

Different from WAVE, MIDI file is non-wave sound file. It stores instruction rather than data. The format of MIDI file that has complication definitions and special encoding rules comes from IEF. MIDI file is composed of two blocks, that is, the file header Mthd and sound channel block Mtrk.

(5) WMA format

WMA format isan audio format that achieves a high compression ratio of 1:18 by reducing data traffic but maintaining sound quality. In addition, DRM scheme can be used to prevent copying, limit insertion time, and number of times, etc, so as to prevent piracy effectively.

WMA format comes from Microsoft Company and the sound quality is better than MP3 format. It supports audio stream technique and is suitable for playing online.

(6) RealAudio format

RealAudio format is the audio formatthat is mainly used for real-time transmission on low-speed networks. It uses 14.4kbps of network bandwidth at its lowest point and is suitable for playing online.

The file format of Real is mainly the followings: RA(RealAudio), RM(RealMedia, RealAudio G2), RMX(RealAudio Secured). These formats are characterized by the ability to change the quality of sound depending on network bandwidth, enabling better sound quality for listeners having better bandwidth while ensuring that most people hear sound smoothly.

Section 2 Sound Recording and Noise Reduction

GoldWave is the audio tool for sound editing, playing, recording, and converting, which has small volume but good function. There are quite a few audio files that can be opened by GoldWave: WAV, OGG, VOC, IFF, AIF, AFC, AU, SND, MP3, MAT, DWD, SMP, VOX, SDS, AVI, MOV, etc. Sound can also be extracted from other video files such as CD, VCD, or DVD. GoldWave can not only trim, stitch the sound arbitrarily, but also process the sound into thousands of strange tones and other masterpieces by using its Doppler, echo, reverb, noise reduction, change of tone, and other functions.

Double-click "GoldWave. exe" and open it, as shown in Figure 3-5. From Figure 3-5, we can see GoldWave starts two panels. The big one is editor and the small one is monitor. The editor completes various edits to the sound waveform, and the monitor controls recording, playback, and some setup operations.

Figure 3-5 Interface of GoldWave

I. Sound Recording

Step 1. Click on the 【File】 menu, select the 【New】 menu item, open the 【New Sound】 dialog box, as shown in Figure 3-6. In this dialog box, Number of channels can select Mono or Stereo. If there is only one microphone recording, then the recording is mono. Here in order to see the functions in the editing area, we choose "2(stereo)"; The default in Sampling rate is "44100", the alternative sampling rate has many options, which can be selected as needed, and Initial file length is the length of the new sound file, that is, the number of time. The format of input value is HH:MM:SS. T. HH stands for hour, MM for minute, and SS for second. The colon is the dividing line. If there is no colon, the number represents seconds; if there is one colon, the front one is minutes and the latter one is seconds. If there are two colons, then the front is hours. As shown in Figure 3-6, the length "1:0" represents one minute. After the setting steps are completed, Click the 【OK】 button and a newly built sound waveform appears in the editing area. As we can see in Figure 3-7, now it is "silent".

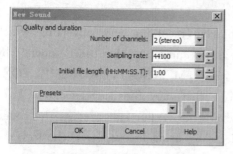
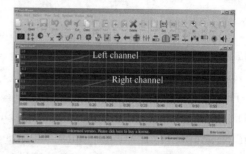

Figure 3-6　New Sound Dialog box　　　　　Figure 3-7　Interface of Waveform

Step 2. Clickthe red 【Recording】 button on the monitor, as shown in Figure 3-8. Then speak to the microphone, a recorded sound waveform appears in the editing area. See Figure 3-9.

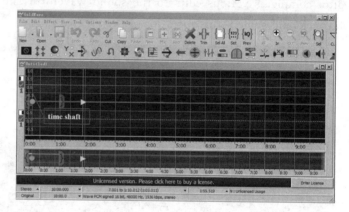

Figure 3-8　Monitor　　　　　Figure 3-9　User's Interface in Recording

Step 3. When the recording is complete, click the 【Play】 button on the Control, as shown in Figure 3-8. Try to listen to the recorded sound.

Step 4. Click on the 【File】 menu and select the 【Save】 menu item. Open the Save Sound As dialog box, as shown in Figure 3-10. Select the file storage path, enter the file name, and select "wav" or "mp3" in Save as type below and click the 【Save】 button to save the sound file.

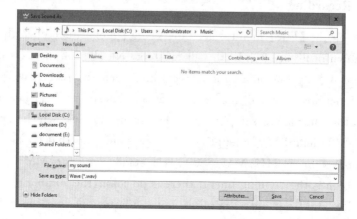

Figure 3-10　Dialogue Box of Save Sound As

> Tips:
>
> 1. When you create a new audio file, make sure the preset time is longer than the required recording time, as it will be easier to intercept if time is longer than not enough.
>
> 2. The default function of the green and yellow play keys on the control is the same, both of the two ones are the waveforms selected for playback, and you can set the contents of the two playback keys in the Control Properties menu item of the Options menu.

II. Sound Editing

During the recording of audio files, we may have some errors, at this time we should pause, and then repeat the wrong sentence, try not to stop recording. After recording is complete, we can delete the wrong place so as to keep the overall consistency of recording of audio. It is very convenient to use GoldWave to edit the sound. The procedures are as followings:

Step 1. Click the 【Options】 menu and select the 【Window】 menu item, as shown in Figure 3-11. Open the 【Window Options】 dialog box and choose the "select left and right mouse button selection method" below, as shown in Figure 3-12, then click the 【OK】 button. After it is completed, click the left mouse button and determine the start point of time of the selected sound clip. Click the right mouse button to determine the finish point of time of the selected sound clip.

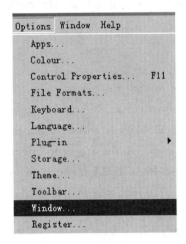

Figure 3-11 Window Menu Item

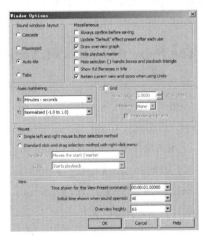

Figure 3-12 Dialogue Box of Windows Options

Step 2. In the control, press the green play button to play the recorded audio file, find the starting position of the sound clip that needs to be deleted, then press the pause button, click the left mouse button where the timeline stops, and determine the starting point of the selected sound clip, as shown in Figure 3-13. Then press the green play button to try, and then press the green play button to continue listening, pause at the end of the sound clip that needs to be deleted, and right-click at the stop on the timeline to complete the selection of the sound clip, with the selected part of the background being blue and the unchecked part of the background being black. See Figure 3-14.

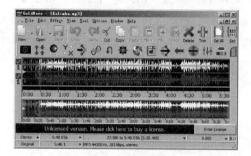 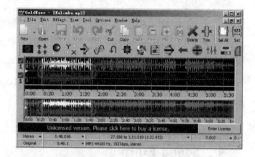

Figure 3-13　Determine the Starting Point of the Selected Sound Clip　　　Figure 3-14　Determine the Finishing Point of the Selected Sound Clip

Step 3. Click the 【Delete】 button in the Toolbar, as shown in Figure 3-15. Delete the selected part and try to listen to the effect. If the sound is very smooth, then the clip of sound is done.

Figure 3-15　Delete Button

Step 4. Repeat the step 2 and step 3. Delete all incorrect sound clips from the recorded audio file and complete the clip for the entire audio file.

Ⅲ. The Noise Reduction of Audio File

Subject to conditions such as recording equipment, the sound we record may contain certain amount of noise. See Figure 3-16. The parts in the yellow rectangle box are noise.

It is very difficult to do the noise reduction. However, the GoldWave software can reduce the effect from the noise on sound quality to a low level. The procedures are as followings:

Step 1. Click on the 【Effect】 menu and select the 【Noise Reduction】 menu item in the cascading menu of the 【Filter】 menu item. See Figure 3-17. Open the Noise Reduction dialogue box, as shown in Figure 3-18.

Step 2. Keep the default value of the panel. Click the 【OK】 button. After it is completed, the sound audio wave is shown in Figure 3-19. The noise decreases sharply as it can be seen that the silent part is approximately a straight line. Play it again, and there is almost no noise, but the sound doesn't change a lot.

The default parameters are removing the noise based on a common device of noise model, but after all, the noise is generated because of different reasons. The factors such as the current environment of each recording, the use of equipment, software, etc. , result in different noise, so GoldWave also provides a method that sample can be selected from the

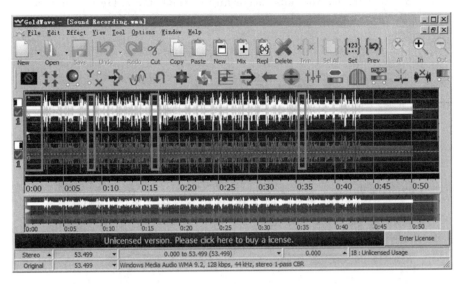

Figure 3-16　Sound Noise in the Waveform

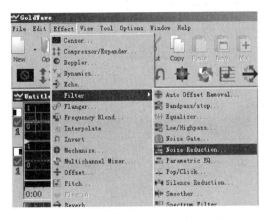

Figure 3-17　Noise Reduction Menu Item

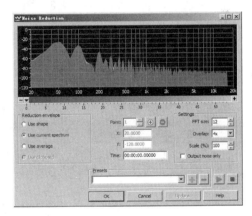

Figure 3-18　Noise Reduction Dialogue Box

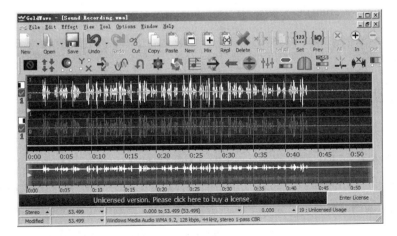

Figure 3-19　Waveform after Sound Reduction

recording file, and then reduce the noise according to the sample. Here's how:

Step 1. Select a waveform with no voice and only noise from the waveform file. See Figure 3-20.

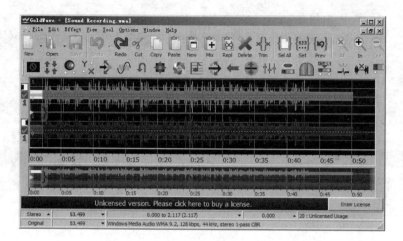

Figure 3-20　Select a part of waveform of noise

Step 2. Try to listen after selection and make sure that there is no sound in the selected part. Click the 【Copy】 button in the Toolbar, as shown in Figure 3-21. Copy the selected part, which is "the sample of noise".

Figure 3-21　Copy Button

Step 3. Click the Select All button in the Toolbar, as shown in Figure 3-22. Select the whole waveform of sound file.

Figure 3-22　Select All Button

Step 4. Click the 【Effect】 menu and select the 【Noise Reduction】 menu in the cascading menu of the 【Filter】 menu item. See Figure 3-17. Open the Noise Reduction dialogue box. Select Use clipboard single option in the Reduction envelop option group at the bottom left of the dialog box, as shown in Figure 3-23, and then click 【OK】 to confirm.

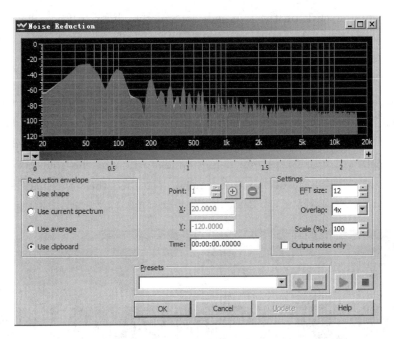

Figure 3-23 Noise Reduction by Using Noise Sample

Section 3   Sound Editing and Synthesis

I. Adjustment of Sound

The volume can be adjusted by using the volume bar in the computer, the speaker or the volume potentiometer in the headset. In addition, some multimedia players also have volume adjustment. However, the various ways mentioned above doesn't change the waveform amplitude of the original sound file, only change the volume of sound played out. Here we are going to change the waveform amplitude of original sound file. The operation steps are as follows.

Step 1. Click the 【Effect】 menu. Choose 【Change Volume】 in the cascading menu of 【Volume】. Open Change Volume dialogue box, as shown in Figure 3-24.

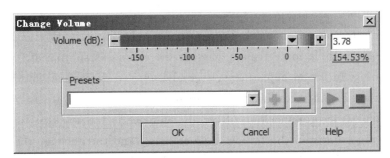

Figure 3-24 Dialogue Box of Change Volume

Step 2. Drag the slider to the left to decrease the volume, and drag the slider to the right to increase the volume; click on the two ends + - of the slider to fine-tune, or enter values directly in the text box on the right side of the slider to adjust.

Step 3. After adjustment, click on the green play button to listen. If it is appropriate, click the 【OK】 button and finish the adjustment of volume.

It should be noted that the computer records a certain range of sound waveform amplitude, that is to say, there is a limit on the maximum value. If the amplitude exceeds this limit, then the computer will only record the maximum value, thus the exceeded part will be intercepted, which will cause a certain degree of distortion of the sound. We call it "top-cutting distortion". The cause of this phenomenon is called the "overload" of volume. See Figure 3-25.

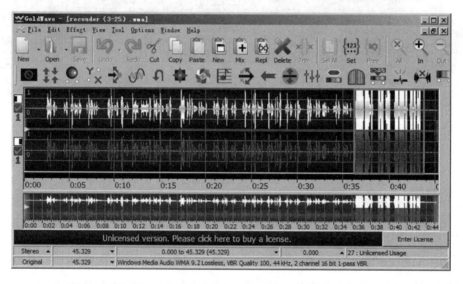

Figure 3-25 Normal Sound Volume and Overload Sound Volume

Tips:

1. The top-cutting distortion can't be deleted by any method, so the volume should be small instead of large in adjustment. In Figure 3-25, the upper and lower lines of the edit box are the limiting lines. It is marked as "±1" and the middle line is "0", meaning "no sound". In sound recording, the wave range should be controlled between 50%-80% as shown in the left part of Figure 3-25.

2. For different sound playback devices, the sound volume is different. Therefore, it is not completely credible to determine whether the volume is appropriate or not only by subjective hearing. In sound adjustment, it is advisable to determine whether the sound size is appropriate or not according to the sound wave.

3. Sometimes the pronunciation of certain sentence is too weak or too strong, then we can select the sound clip that needs to be adjusted, and then change the volume, thus our volume adjustment only works on the selected part.

4. The method of selecting first and then adjusting can achieve local adjustment of sound, which is valid to all effect adjustment.

5. When selecting only one sentence or even one word, you can use the Zoom In button on the toolbar to zoom in on the sound's time scale to make the selection more accurate. When you don't need to zoom in on the display, click the Zoom Out button to make the waveform zoom out.

II. Sound Synthesis

The synthesis of sound is mainly used for the combination of two sounds so that two sounds can be played at the same time, such as soundtrack recitation. The steps are as follows:

Step 1. Click the 【Files】 menu, select the 【Open】 menu item, as shown in Figure 3-26. Open the dialogue box of Open Sound and select the background music file, as shown in Figure 3-27. Click 【Open】 button and open the background sound file.

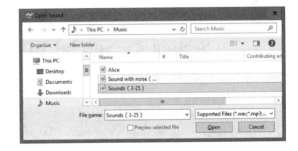

Figure 3-26 Open Menu Item   Figure 3-27 Dialogue Box of Open Sound

Step 2. Select the sound clip or the complete one used for background music. Click the 【Copy】 button in the Toolbar. See Figure 3-28.

Figure 3-28 Copy button

Step3. Same with step 1. Open the sound file.

Step 4. Click the 【Mix】 button, as shown in Figure3-29. Open the dialogue box of Mix, press the green 【Play】 button and listen to the mix effect. Drag the slider of Volume to adjust the volume of the background music, as shown in Figure 3-30. Generally the volume of the background music should be lower than that of the sound. If it is done, click the 【OK】 button to stop.

Figure 3-29 Mix Button

Figure 3-30 Mix Dialogue Box

Section 4 Special Effect Processing of Sound

Ⅰ. Adding Echo

Echo, as the name implies, means that the sound is heard after a certain amount of time and then returned to be heard by us, just as people cry in the wilderness in the face of the mountain. It is widely adopted in many film and television clips. The echo effect by using GoldWave is simple to do, as follows:

Step 1. Click the 【Effect】 menu and select the 【Echo】 menu item, as shown in Figure 3-31. Open the Echo dialogue box as shown in Figure 3-32.

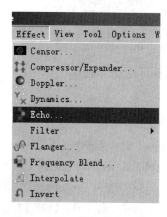

Figure 3-31 Echo Menu Item

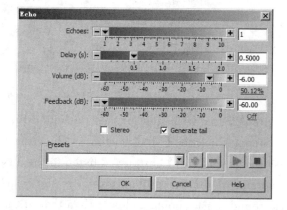

Figure 3-32 Echo Dialogue Box

Step 2. Adjust the parameters. The parameter Echoes is used to set the number of echo repeats. The more times the number of repeats, the more obvious the effect is. The parameter Delay is used to set the time of delay, that is, the time difference between the acoustic sound and the echo. Theoretically, it is the time spent for twice of the distance between sound source and the reflector. The higher the value, the longer the sound lasts. The parameter Volume is used to set the size of the echo volume, which should not be too large, otherwise the echo effect will appear unreal. The parameter Feedback is used to set the volume of the reply back wave, and adjust the depth of the wave. When it is done, press the green 【Play】 key in the window to listen and adjust the unsatisfactory places again until you're satisfied, click the 【OK】 button to complete the setup.

Ⅱ. Adding Reverb Effect

Reverberating effects are used in many songs and audios to increase the appeal of sound. The principles of reverberation and echo are very similar. The difference lies in that reverberation simulates the sound effect of the concert hall, cinema and other closed space while the echo is from all sides, thus reverberating effect can be understood as the overlying echo effect from all sides. It is easier to add reverb effect by using GoldWave. The steps are as follows:

Step 1. Click the 【Effect】 menu and select 【Reverb】 menu item, as shown in Figure 3-33. Open the Reverb dialogue box. See Figure 3-34.

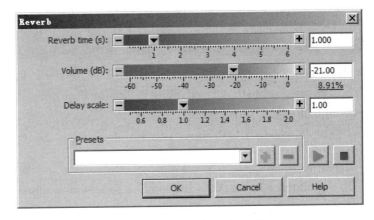

Figure 3-33　Reverb Menu Item　　　　　Figure 3-34　Reverb Dialogue Box

Step 2. Adjust the parameters. The parameter Reverb Time is used to set the length of the reverb sound, generally using the default value of 1.0; the parameter Volume is used to set the size of the reverb sound. The lower the volume, the quieter the reverb. The parameter Delay Scale is used to fine-tune the reverb effect, typically taking 1.0. When you're done setting up, you can press the green 【Play】 key to listen and adjust the unsatisfactory places again until you're satisfied, click the 【OK】 button to complete the setup.

Ⅲ. Adjustment of Pitch

Thepitch adjustment we make is to change the vibration "frequency" of the sound.

Generally speaking, women's vocal cord is tight and thin, and the pitch is higher while men's vocal cord is loose and thick, and the pitch is low. However, with the tuning function, we can turn a female high pitch into a male bass or a male bass into a female high pitch, as follows:

Step1. Click the 【Effect】 menu and select 【Pitch】 menu item. Open Pitch dialogue box as shown in Figure 3-35.

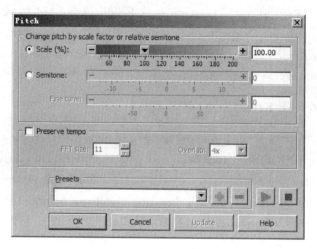

Figure 3-35　Pitch adjustment Dialogue Box

Step 2. For speech adjustments, generally we select Scale. Drag slider to reduce or increase the pitch proportionally, or enter numbers in the text box on the right. For example, enter 110 to indicate that the frequency of the sound becomes 110%; for the adjustment of music, we select Semitone. After selecting the Semitone item, click the - on left-hand side each time and click the + on the right-hand side to do the rise and fall within half of the scale. You can make adjustments in the next line of Fine tune if you want to do the rise and fall within half of the scale.

Step 3. Select the Preserve tempo item to keep the length of the audio file unchanged.

Tips:
1. Pitch adjustment is the same as the volume adjustment, or you can choose to have a sound clip or even a word tone to change.

2. When the pitch changes a large amount, the volume of the sound will also change accordingly. Generally we need to adjust the volume depending on the situation after we adjust the pitch.

Chapter 4

Digital Video Processing

A video is a set of images that change continuously over time. When the continuous image changes more than 24 pictures per second, according to the principle of visual retention, the human eyes can not distinguish a single static picture, thus a smooth continuous visual effect will be produced. This continuous change of the picture group is called video or dynamic images.

Video is obtained directly from the real world through video camera, which can make people recognize and understand the meaning of multimedia information. Video can be categorized as analog video and digital video.

Section 1 Fundamental Knowledge of Video

I. Analog Video and Digital Video

The term video comes from the Latin term "I can see" and usually refers to different kinds of motion pictures, also known as filmsand videos. It generally refers to various techniques of processing a series of static images by capturing, recording, storing, and transmitting as electrical signal. According to the way of storage and processing, videos can be divided into analog video and digital video two categories.

1. Analog Video

Analog Video belongs to the traditional category of TV video signal, which means that each frame image is a real image signal of a natural scene obtained in real time. Analog video signal is based on analog technology and international standards for image display to produce video footage, with advantages of low cost, good reductive property. Video screen often gives people an immersive feeling. The disadvantage is that no matter how good the recorded image signal is, the quality of the signal and picture will be significantly reduced after a long period of storage or multiple copies.

TV signal is an important source of information for video processing. The standard for television signals is also known as the system of television. At present, the television standards of different countries are different. The main differences lie in different refresh speeds, color coding systems and transmission frequencies. At present, the world's most commonly used analog broadcast video standards (systems) are the PAL systems used by China and Europe, the NTSC system used by the United States and Japan, the SECAM system used by France and other countries.

(1) NTSC System

NTSC standard was established by the National Television Standard Committee in 1952. Its basic content is as follows: the frame of the video signal consists of 525 horizontal scanning lines which adopt the mode of interlaced scanning that every 1/30 seconds do the refresh in the surface of the kinescope and each frame screen is completed by two scans in which draws a field each time within 1/60 seconds and two fields constitute a frame. NTSC is used in the United States, Canada, Mexico, Japan, and many other countries.

(2) PAL System

PAL (Phase Alternate Lock) standard is compatible television system made in Germany in 1962, mainly used in Australia, South Africa, China, South America, and most European countries.

(3) SECAM System

SECAM standard is the abbreviation of Sequential Color and Memory. It is mainly used in France and Eastern Europe. It is a 625-wire, 50 Hz system.

Analog video signals mainly include brightness signals, chromaticity signals, composite synchronization signals, and accompaniment signals. The YAV model is used to represent color images in PAL color TV system. Among them, Y represents brightness, U and V represent color difference, and they are the two components that make up colors. Similarly, the YIQ model is used in the NTSC color TV system, where Y represents brightness, and I and Q are two colored parts. The importance of the YUV notation is that its brightness signal (Y) and chroma signal (U, V) are independent of each other. That is, the black-and-white grayscale map of the Y signal components and the other two monochrome maps made up of U and V signals are independent of each other. As Y, U, and V are independent, these monochrome maps can be encoded separately.

2. Digital Video

Digital video is relative to analog signals and refers to video recorded in digital form. Digital video has different ways of producing, storing, and playing. A video acquisition card can be used to realize the conversion of A/D (analog/digital) of analog video signal, which is the process of video capture (or acquisition). The converted signal is digitally compressed stored into the computer hard disk. Thus the analog video is converted into digital video.

Compared with analog video, the digital video has the following features:

①Digital video can be reproduced countless times without distortion.

②Digital video is easy to store for long periods of time without any quality reduction.

③Digital video can be done with NLE(non-linear editing), and stunts can be added.

④There is large amount of digital video data. In the process of storage and transmission, compressed encoding must be done.

Ⅱ. Linear Editing and Non-linear Editing

1. Linear editing

Linear editing is the traditional way of editing video. The sequence of video signals is recorded on tape. When editing a video, the editor selects a suitable piece of footage by playing the tape on a video recorder and records it on a tape in the recorder, and then looks for the video footage in order and records it. Do this over and over until all the appropriate footage is recorded in the full order required by the program. This sequential way of editing video is called linear editing.

2. Non-linear editing

Nonlinear video editing is for digital video files. Video post-editing production can be done in the computer's software editing environment. Random access to any part of the original footage, modification and processing can be achieved.

This way of editing non-sequential structures is called nonlinear editing.

Nonlinear editing has the following features:

①Nonlinear editing footage is stored in the form of digital signals on the computer hard disk, which is user-friendly and can complete the fast search, accurate positioning, and the control of the quality of image.

②Nonlinear editing has powerful editing capabilities. A complete nonlinear editing system often integrates recording, editing, stunts, subtitles, animation, and other functions, which is unmatched by linear editing.

③The investment on nonlinear editing system is relatively small in that equipment maintenance, repair and operating costs are much lower than linear editing.

The features of nonlinear video editing have made it the main way of TV program editing.

Ⅲ. Digitalization of Audio Signal

There are two main ways to obtain digital video information: one is the use of digital cameras to capture scenes to directly obtain distortion-free digital video; the other is to convert analog video into digital video by using the video acquisition card, and save it in the format of digital video files.

A digital video acquisition system consists of three parts: a highly configured multimedia computer system(MPC), a video acquisition card, and a video source. See Figure 4-1.

1. Function of Video Acquisition Card

On the computer, the video acquisition card can receive analog video signals from video inputs(video recorders, cameras and other video sources). The signal is collected, quantified and made into a digital signal, and then it is compressed and encoded into a digital video

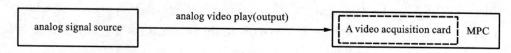

Figure 4-1 Acquisition System of Digital Video

sequence. Most of the video capture cards have the feature of hardware compression. When capturing a video signal, first, compress the video signal on the card; then transmit the compressed video data to the host computer via the PCI interface. The common video acquisition card stores the digital video into AVI files by using in-frame compression method while the high-end video acquisition card can also directly compress the collected digital video data into MPEG-1 format files in real-time.

As analog video inputs can provide an uninterrupted source of information, the video acquisition card captures each frame image in the analog video sequence and passes the data into the computer system before capturing the next frame image. Therefore the key to real-time acquisition is the processing time required for each frame. If the processing time of each frame of video image exceeds the time between the adjacent two frames, the loss of data occurs, i.e. the loss of frames. Acquisition cards compress the acquired video sequence before storing on the hard disk. That is, the acquisition and compression of the video sequence is done together, saving the trouble of compressing again.

2. The Working Principle of Video Acquisition Card

The structure of videoacquisition card is shown in Figure 4-2. Multi-channel video input receives the video input signal. The video signal source first converts the analog signal into a digital signal through the A/D (analog/digital) converter, and then it is trimmed, scale-changed by the video acquisition controller before being compressed into the frame memorizer. When outputting the analog video, the contents of the frame memorizer are converted into analog signals by a D/A (digital/analog) converter and outputted to a television or video recorder.

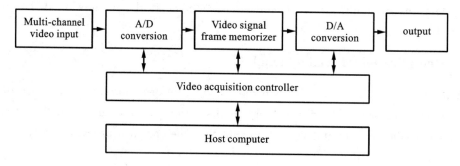

Figure 4-2 Structure of Video Acquisition Card

3. Digital Video Output

The output of digital video is the inverse process of digital video acquisition, that is, the digital video signal is converted into analog video signal output to the television for display, or output to the recorder recorded in the tape. Similar to video acquisition, this requires the use of specialized equipment to extract digital video and D/A transformation to complete the conversion of digital data to analog signals. Depending on the application and needs, there are various kinds of device for conversion. The high-end video acquisition card, which combines analog video acquisition and output, is inserted in the expansion slot of the PC. It can be connected to a more specialized video recorder to provide high-quality analog video signal acquisition and output. This device can be used for professional video acquisition, editing and output.

In addition, there is another device called TV Coder, the function of which is to convert everything displayed on the computer monitor into an analog video signal and output it to a television or video recorder. The function of this device is limited and suitable for ordinary multimedia applications.

Ⅳ. Digital Video File Format

1. Ordinary Video File Format

(1) AVI format

AVI(Audio Video Interleaved) is a digital video file format of audio and video cross-recording, and the motion image and accompaniment data are stored alternately. This interweaving way of audio and video is similar to traditional movies in that there are the frame sequence with image information and display of accompanying sound track in the movies.

The AVI file structure not only solves the problem of audio and video synchronization, but also has the feature of being open and universal. It works in any Windows environment and has the ability to extend it. Users can develop their own AVI video files, which can be used at any time in a Windows environment.

AVI adopts the in-frame loss compression. It can be re-edited and processed with ordinary video editing software(e. g. Adobe Premiere). The advantage of this file format is that the quality of image is the best and it can be used across platforms.

(2) MPEG Format

The suffix of MPEG(Moving Picture Expert Group) format is mpeg. The VCD, SVCD, and DVD used in the household are MPEG format files.

The full-screen active video standard file can be generated by using the MPEG algorithm method to compress the full-motion video image. MPEG format file can simultaneously play full-motion video images and CD music accompaniment at a rate of 25 frames per second(or 30 frames per second) at a resolution of 1024 × 786, and its file size is only 1/6 of that of the AVI file. MPEG-2 compression technology uses variable rate (Variable Bit Rate, VBR) technology, which can change the data transmission rate in real time according to the complexity of the dynamic picture in order to obtain better encoding results. This technique is used in DVDs currently.

The average compression rate of MPEG is 50:1 and the maxim rate can reach to 200:1, which is very high. Meanwhile, the quality of image and sound is very good. The standard of MPEG include MPEG video, MPEG audio and system. MP3 audio file is the typical application of MPEG audios while the VCD, SVCD and DVD are the new consumer electronics produced by using MPEG technology.

(3) MOV Format

MOV(Movie digital Video technology) is the video file format developed by American Apple company. The default player is QuickTime Player, which has a high compression ratio and better video clarity, and can be used across platforms.

2. Network video file format(streaming format)

(1) RM format

As the contemporary main stream network video file format, RM is the streaming media file format developed by the RealNetworks company. The audio and video compression standard made by RealNetworks is called Real Media and the player is Real Player.

(2) ASF Format

ASF(Advanced Streaming Format) is the streaming media format of Microsoft company in earlier stage, which adopts the MPEG-4 compression algorithm. It is the video file format can be viewed in real time on the Internet.

(3) WMV Format

WMV(Windows Media Video) format is the video file format with independent encoding method developed by Microsoft company, which is one of the most widely used streaming video format currently.

Section 2　Basic Editing Procedures of Video Footage

Video Studio is a powerful video editing software with function of video capturing and editing produced by Corel Company in Canada. It supports MV, DV, and other devices to obtain video files, providing a lot of editing functions and effects, and can derive a variety of common video formats. Double-click the Corel Video Studio icon to open the audio-visual editing interface. The functional zone is shown in Figure 4-3.

Ⅰ. Obtaining Footage

Video Studio obtains footages by capturing it from a video device, importing it from DVD/VCD or from mobile device. Of course, it can also import footage that already exists on the hard drive directly.

1. Obtaining Footage

Obtaining or importing footage from the exterior device. The procedures are as follows:

In the Video Studio interface, click on the 【Capture】 item and choose the way of loading footage according to the source of footage in the property panel on the right. See Figure4-4. For example, Capture Video enables users to record video footage directly from the computer's camera, and other options require connecting the appropriate external device to the computer for footage capture.

Chapter 4　Digital Video Processing

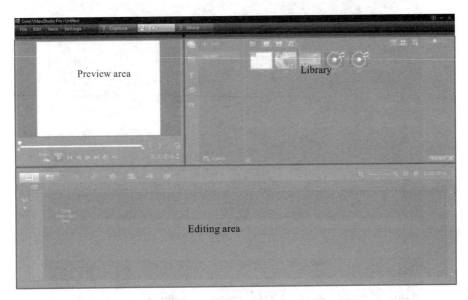

Figure 4-3　The Functional Zone of Video Studio

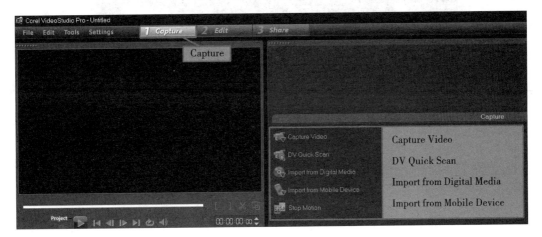

Figure 4-4　Window of Capture

2. Importing Footage

Import footage that already exists on the hard drive, as followings:

Step 1. In the interface, choose 【Edit】 item, as shown in Figure 4-5.

Step 2. In a column of buttons on the left side of the library, click the 【Media】 button, as shown in Figure4-6. To import the media files into the media library, click the 【Import media files】 button, then open the Browse media files dialogue box, find the media file you want to import in that window, hold down Ctrl if you need to select more than one, click the file, you can add the selected file until you select all the files you need. See Figure4-7. Then click on 【Open】 button to import the selected media files in the media library.

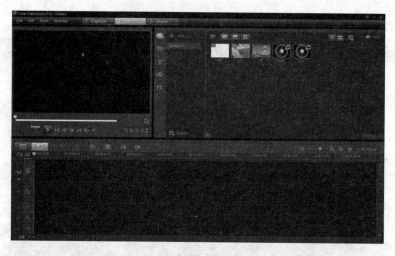

Figure 4-5 Interface of Video Studio

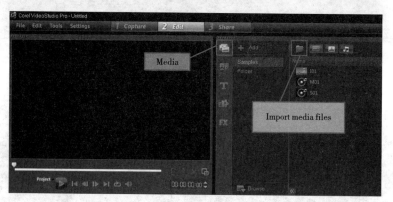

Figure 4-6 Loading Footage

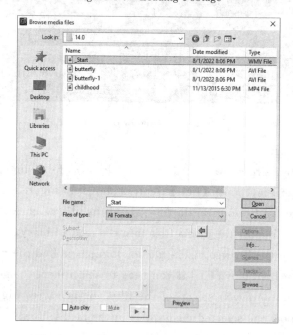

Figure 4-7 Browse media files Dialogue Box

> Tips:
> Importing audio footage requires clicking the Audio button in a column of buttons to the left of the library, and then clicking the Add button to import the audio file into the library.

II. Editing the Footage

Once you've finished importing all the footage you need for video editing, you're ready to make basic edits to the footage.

1. Loading Footage

Step 1. Click the 【Timeline View】 button at the left-hand position above the edit area to enter the timeline view mode, as shown in Figure 4-8, and the default timeline view panel is top-down, 【Video Track】【Overlap Track】【Title Track】【Sound Track】【Music Track】.

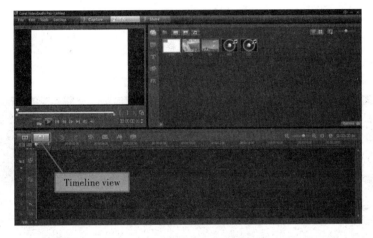

Figure 4-8 Timeline View Button

Step 2. Drag the main content that needs to be displayed to the editing area. Then drag the video and photos to the 【Video Track】 in turn, the voice to the 【Sound Track】, the music to the 【Music Track】, as shown in Figure 4-9

Figure 4-9 Importing Footage

2. Video Editing

(1) The Basic Clipping of Video

Step 1. Select the video file that needs to beclipped, and the selected file will be framed with a yellow rectangle, as shown in Figure 4-10.

Figure 4-10　Select the Video File

Step 2. Click the 【Options】 button in the lower right corner of the 【Footage Gallery】, as shown in Figure4-11. Click to open the 【Video Options】 panel. Reduce the length of the video in the input box 【Video Interval】, as shown in Figure 4-12. This is equivalent to keeping the content from the beginning of the video to the set point in time, with the content later deleted.

Figure 4-11　Options Button

Figure 4-12　Video Options Panel

Step 3. Click on the 【Play】 button in the Preview area, as shown in Figure 4-13 to preview the video file. The white time pointer in the edit area moves back, as shown in Figure 4-14. After playing to the place where you need to clip, click the 【Pause】 button, the time pointer stays in the current position, click the 【[】 button in the preview area to delete the video to the left of the time pointer. Click the 【]】 button on the right side of the time pointer to delete the video on the right side of the time pointer, and click the small scissors button on the right side of the 【]】 to "cut" the video file from the time pointer position and split the video file into two videos

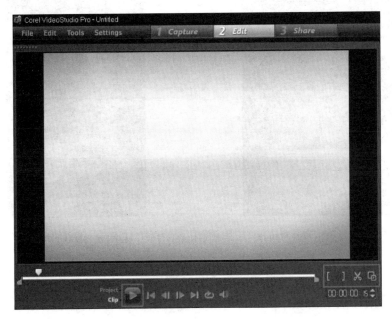

Figure 4-13 Play Button and Edit Buttons in the Preview Window

Figure 4-14 Time Pointer

(2) TheAudio-visual Separation of the Video

Whetherit's captured video footage or imported video footage, the picture and sound are synthesized together. When we need to adjust only the sound or picture in the video, we need to first separate the sound and picture of the video, that is the sound and picture separation of the video. Here's how:

Drag the video footage to the 【Video Track】, and right-click on the video footage, select the 【Split Audio】 option, as shown in Figure 4-15. Then the picture is still on the 【Video Track】, but there is no sound. The sound icon in the lower right corner will be muted, and there will be an audio file on the 【Sound Track】, as shown in Figure 4-16. Now, sound or picture can be individually adjusted, but special attention should be paid to ensure that the audio and picture of the video work is synchronized.

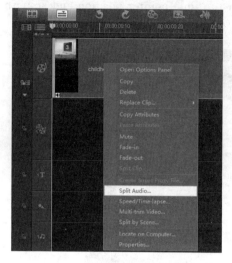

Figure 4-15　Split Video and Sound　　　　Figure 4-16　Footage Added in the Sound Track

(3) Picture Editing

Step 1. Check the photo file that needs to be edited, the selected file will also be boxed in a yellow rectangle, and the time pointer will automatically move to the starting point of the file. See Figure 4-17.

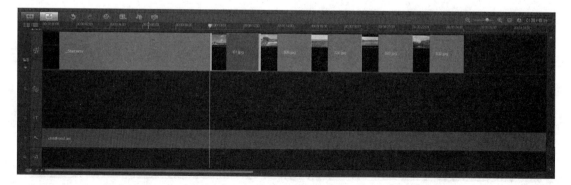

Figure 4-17　Select Picture

Step 2. 【Options】 button in the lower right corner of the 【Footage Gallery】 to open the 【Photo Options】 panel. Adjust the length of stay of the photo in the 【Photo Interval】 input box, as shown in Figure 4-18. The default length of time is 3 seconds, you can turn up or down.

Figure 4-18　Photo Options Panel

> Tips:
> 　　1. The audio editing methods mentioned above are the two of audioediting method. You can choose either of them, or you can use both alternately.
> 　　2. The time pointer is also known as the "flying shuttle bar".
> 　　3. The method of audio editing is similar to video clips.
> 　　4. The footage on the Video Track must start from 0, and there is no time interval between footages. When deleting certain footage or the front part of the video file, the content on the right side of it will automatically move forward, connecting with the footage on the left side of the deleted content. There should be no "empty part".
> 　　5. Except for the Video track, the footage on other tracks such as Music track can start from any moment. Therefore, when deleting the content on the left side, the content on the right side will not automatically move forward. One should adjust the start point of the footage as needed.
> 　　6. When editing the footage, pay attention to the total length of time of each track, and time of audio and music footage generally should not exceed the video track and title track of the footage time, so as to avoid the "black screen phenomenon", screen with sound but no pictures.

Ⅲ. Adding Transition

　　Directly arranging thefootage on the video track, in preview we will find that the direct connection of the footage is very rigid. In order to ease this rigid transition, we can add the transition effect directly between two footages, add the specific steps are as follows.

Step 1. Click on the 【Transition】 button in a column of buttons on the left side of the library. Select 【All】 or a set of required transition effects in the drop-down menu as shown in Figure4-19.

Step 2. Click on certain effect in the footage gallery to preview. After selecting the needed effect, drag it to the place that is between the two footage, as shown in Figure 4-20. The time pointer automatically stays at the starting point of the transition effect, and the transition effect is selected.

Step 3. Click the 【Options】 button in the lower right corner of the 【Footage Gallery】 to open the 【Transition Effect Options】 panel, which has different options panel parameters for different transition effects. See Figure 4-21, Figure 4-22. However, it is possible to adjust the length of the transition time in the 【Interval】 input box, and other parameters can be adjusted to see its effect.

Figure 4-19　Transition Effect List

Figure 4-20　Adding Transition Effect

Figure 4-21　Transition Effect Options

Figure 4-22　Another Transition Effect Options

Section 3　Operation of Special Effect of Video

The special effect of video isn't the necessary element for videos, but the proper use of special effect can make the video more delicate with stronger expressive force.

I . Adding Filter

The filter effect is mainly to add some transformations or decorations to the video or photo to make the picture vivid. Here's how to add filters.

Step 1. Click the 【Filter】 button in a column of buttons to the left of the library, select 【All】 or certain filter effect in the drop-down menu shown in Figure 4-23.

Step 2. Click a filter in the Footage Gallery panel to preview. Select the filter you want, drag it to the video or photo footage, the time pointer automatically stays at the beginning of the footage, and a small black square mark appears on the footage thumbnail, the footage is selected.

Step 3. Click the 【Options】 button in the lower right corner of the 【Footage Gallery】 to open the Attribute tab of the Options panel, as shown in Figure 4-25, which lists all the filter effects added to the currently selected footage. If you need to add more than one filter effect, you need to click in the check box in front of 【Replace Last Filter】 to uncheck. See Figure 4-26.

Step 4. Select any of the filters in the list of added filter effects, and click the drop-down list arrow below to open a list of preset effects for that filter, as shown in Figure 4-27. You can also click the 【Customiz Filter】 link to open the filter parameter settings window and set

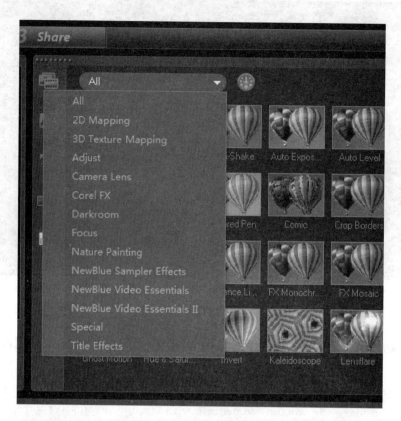

Figure 4-23　Filter Effect List

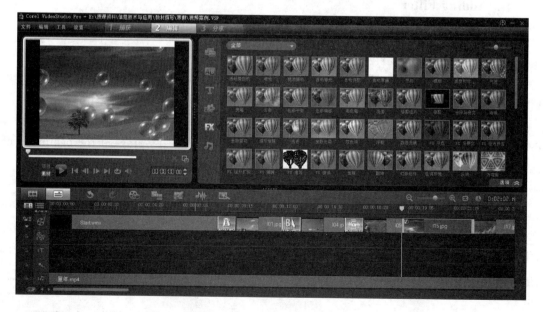

Figure 4-24　Adding Filter Effect

Figure 4-25 Footage Attribute Tab

Figure 4-26 Add More Filter Effect

the filter parameters, as shown in Figure 4-28. Note that different filter effects have different parameters, so the settings interface is different.

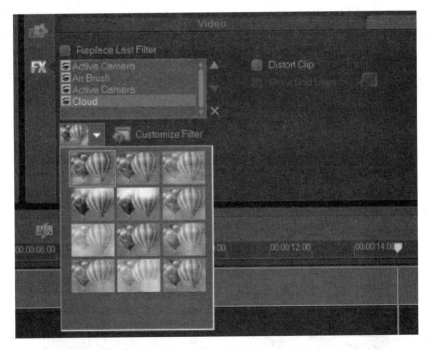

Figure 4-27 Preset Filter Effect List

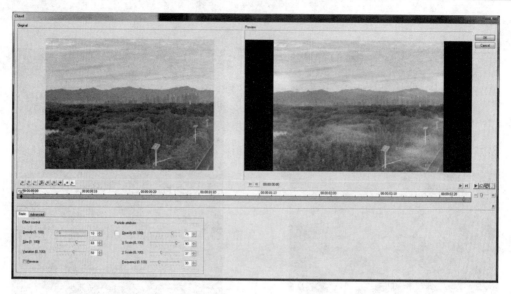

Figure 4-28 Cloud Filter Settings Window

II. Picture-in-picture Effect

Picture-in-picture effect is a common method of video presentation, which is used in many video works. The Video Studio has good support for the picture-in-picture effect. The specific setting method is as follows:

Step 1. Click the 【Play】 button in the 【Preview Area】 to preview the video, and click the 【Pause】 button when you need to add a picture-in-picture to keep the time pointer at the beginning of the picture-in-picture.

Step 2. Drag the sub-picture footage to the time pointer position of the 【Overlay Track】, as shown in Figure 4-29. There will be 8 yellow control points and 4 green control points on the sub-picture screen in the preview window. The yellow control points will be used to resize the screen and the green control points will be used to adjust the shape of the screen.

Figure 4-29 Add Picture-in-picture

Figure 4-29　Add Picture-in-picture

Step 3. Drag the yellow control point in the preview window to adjust the size of thesub-picture. Drag the green control point to adjust the shape, and drag the animation surface to adjust the position of the sub-picture. See Figure 4-30.

Figure 4-30　Adjust Sub-picture

Step 4. The footage of the sub-picture screen can also be added a filter effect. Click the 【Option】 button in the lower right corner of the 【Library】. You can adjust the number of filters in the filter settings area of the 【Properties】 panel. Click the 【Customiz Filter】 icon to adjust the filter parameters. See Figure 4-31.

Step 5. In the animation area(see Figure 4-31), you can set the entry, exit, and fade-out animations of the sub-picture.

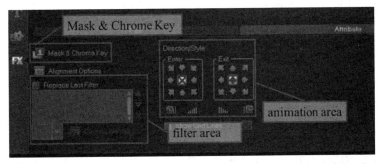

Figure 4-31　Attribute Option Menu of Overlay Footage Option Panel

Step 6. Click the 【Mask & Chroma Key】 link at the top left of the 【Properties】 panel to open the 【Mask and Chroma Key】 panel, where you can check the 【Apply Overlay Option】 check box, select 【Mask Frame】 in the 【Type】 drop-down list, and select a shape in the shape list on the right, as shown in Figure4-32. Then the sub-picture is clipped into the selected shape, as shown in Figure 4-33.

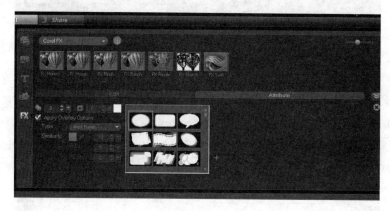

Figure 4-32 Mask and Chroma Key Panel

Figure 4-33 Mask Effect

Ⅲ. Processing of "Blue Curtain"

"Blue Curtain" technology, also known as chromo-key technology, is a processing method often used in film and television works. It is shooting objects in front of a single color background and in the later video processing, distinguishing the foreground and background according to the background color information. Usually the foreground is retained and the background is replaced with a different picture. Then a new video work is synthesized. The Video Studio can provide good support for the blue curtain effect. The specific procedures are as follows.

Step 1. Click the 【Play】 button in the 【Preview Area】 to preview the video, and click the 【Pause】 button when you need to add a foreground through Blue Curtain technology to keep the time pointer at the beginning of the Blue Curtain.

Step 2. Drag footage with a single background color to the time pointer position of the 【Overlay Track】, as shown in Figure 4-34. Drag the yellow control points in the preview window to resize the footage. Drag the footage to adjust the position. See Figure 4-35.

Figure 4-34 Add Blue Curtain Footage

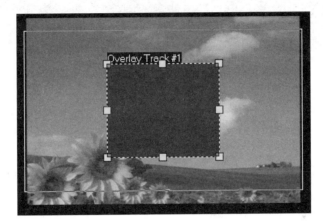

Figure 4-35 Preview Window

Step 3. Click the 【Option】 button in the lower right corner of the 【Library】. Click the 【Mask & Chroma Key】 link in the 【Properties】 panel, as shown in Figure 4-31. Open the 【Mask & Chromo Key】 panel, where the 【Apply Overlay Options】 check box is checked. Select the 【Chroma Key】 in the Type drop-down list and the software will automatically select the background color of the screen. If the selected color is not proper, then click the picker icon, as shown in Figure 4-36. Use the color picker to click in the preview window, and

Figure 4-36 Set the Parameter of Color Key

select the appropriate background color. After the Blue Curtain effect is set, the preview window screen is shown in Figure 4-37.

Figure 4-37 Preview Window of Blue Curtain Effect

Tips:

1. The step 3 to step 6 in picture-in-picture effect are selectable steps. It can be used when needed.

2. You can also add filters and animations to your footage while you're working onthe Blue Curtain.

3. When you need two overlapping effects at the same time, Picture-in-Picture and Blue curtain, click the [Track Manager] button above the video track icon in the edit area as shown in Figure 4-38. Open the overlay Track Manager window, see Figure 4-39. In this window, you can set the tracks that appear in the edit area by checking the box. The sound will give us up to 6 overlapping tracks, title tracks, audio tracks, music tracks up to 2 each.

Figure 4-38 Track Manager Button   Figure 4-39 Track Manager Window

Section 4 Adding Subtitles

Subtitles are a kind of footage that often appears in many video works, and adding subtitles to the sound will be done on its title track, as follows.

Step 1In a column of buttons to the left of the library, click the 【Title】 button, or click the 【Title Track】 icon in the edit area to call up the 【Title Gallery】 panel, and the "Double-click here to add a title" prompt will appear in the preview window. See Figure 4-40.

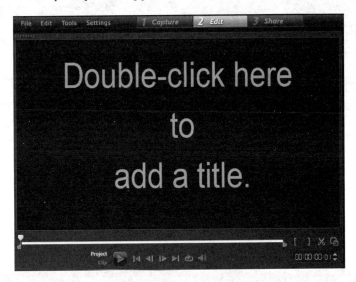

Figure 4-40 Title Editing Prompt

Step 2. Preview the work. Click the 【Pause】 button when you need to add text, the time pointer stays where the text is added, double-click the text prompt in the preview window for text editing. See Figure 4-41.

Figure 4-41 Example of Title Editing

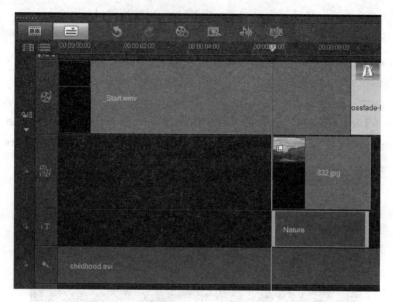

Figure 4-41 Example of Title Editing

Step 3. While editing the text, the 【Title Options】 panel in the library panel opens automatically, as shown in 4-42. In the panel, you can set the length of time, the text, font size, color, and so on. Then click the 【Border/Shadow/Transparency】 option to open the dialogue box 【Border/Shadow/Transparency】, as shown in Figure 4-43, to finish the setting.

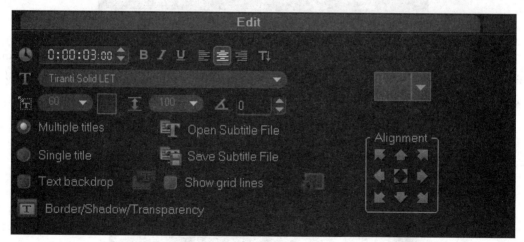

Figure 4-42 Title Options Panel

Step 4. Click the 【Attribute】 tab of the 【Title Options】 panel to animate the text. See Figure 4-44. Check the box in front of 【Apply】, as shown in Figure 4-45. In the drop-down box, select a motion, such as Fly, and then select the animation you want in the list of preset effects below it, as shown in 4-45. You can also click the 【Custom Animation Properties】 button to the right of the pull-down list box, as shown in Figure 4-46, set your own animation parameters in the pop-up dialog box.

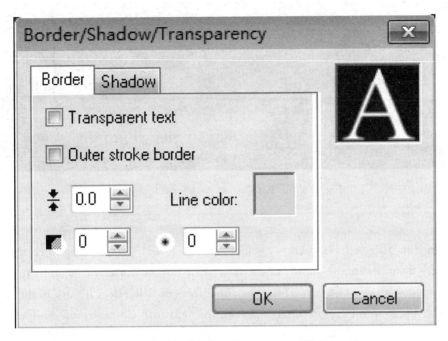

Figure 4-43 Border/Shadow/Transparency Dialogue Box

Figure 4-44 Title Attribute Tab

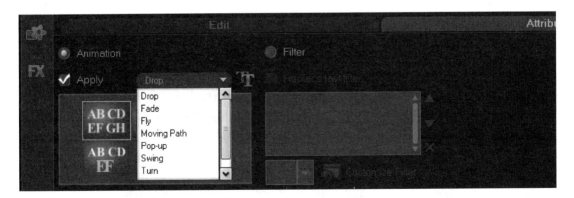

Figure 4-45 Animation Effect List

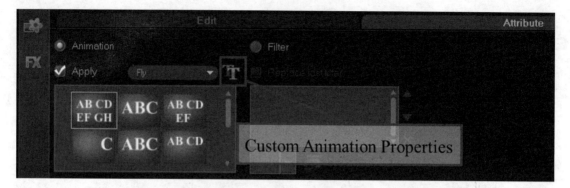

Figure 4-46   Set Custom Animation Properties

> Tips:
> 1. In the title effects library, select a preset text effect, drag directly to the appropriate position in the Title Track, and then double-click the sample text in the preview window to edit the text. However, the effects of title effect library in the software are mainly designed for English font. Change the text into Chinese, and double-click the sample text in the preview window to edit the text. However, many effects of color, animation will be invalidated after changing the text into Chinese.
> 2. The text that needs to be added into the video such as titles, subtitles, staff lists, and so on, needs to be added to the title track.
> 3. Preview the video after adding text animation to ensure that the text on the stage has enough time to stay, so that the audience will see the text clearly. Do not enter the animation immediately after exiting the animation.
> 4. The Video Audio supports the import of subtitle files. In the Edit panel, click Open Subtitle Files, you can open the file selection window, import subtitle files. Its supported subtitle file format is .utf format, which is generally generated by a special subtitle production software, including the appearance time of each subtitle, disappear time and other information.

Section 5   Saving and Output

In the process of making video, it is necessary to save the file in time to avoid the loss of the files. Once you're done, you'll need to export the video to be able to play it with a video player.

I. Saving

The engineering file of Video Studio contains the editing information, such as track, clip point, footage order, subtitles, position, time pointer and so on. It is the editable file which can be opened and edited with Video Studio.

Step 1. Click on the 【File】 menu and choose 【Save】 or 【Save As】 menu item, as shown in Figure 4-47.

Step 2. In the pop-up window, select the file storing path and set the file name, and then click the 【Save】 button. See Figure 4-48.

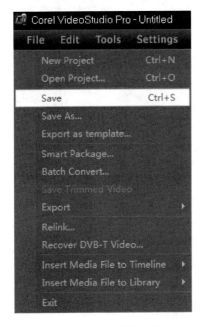

Figure 4-47 Save Menu Item

Figure 4-48 Save As Dialogue Box

II. Video Output

Export video operation is the only way to convert the file into a video work. The exported video work can be separated from the software environment. It can be played using any video player. Here are the steps to export the video.

Step 1. Click on the 【Share】 item. Click 【Create Video File】 in the 【Library Options】 panel on the right. Select a preset video file template as needed in the list box of file templates that pop up, or select the 【Custom】 option to configure the video file parameters yourself. See Figure 4-49.

Step 2. If you use a preset video file template, then open the 【Create Video File】 shown in Figure 4-50. If you use the 【Custom】 option, then the opened create video file dialogue box is shown as Figure 4-51. Click the 【Option】 button to configure parameters such as the video's frame rate, display aspect ratio, compression in the 【General】 and 【Compression】 tabs of the opened 【Video Save Options】 dialogue box. See Figure 4-52.

Step 3. Select the video file storage path. Set the file name, and then click the 【Save】 button.

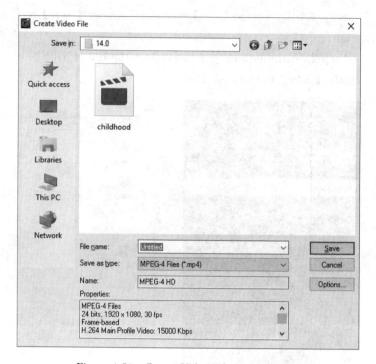

Figure 4-49　Create Video File Option

Figure 4-50　Create Video File Dialogue Box

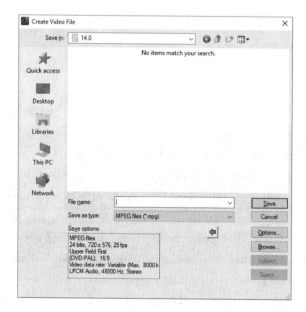

Figure 4-51　Create Video File Dialogue Box of Customized Video Parameter

Figure 4-52　Video Save Options Dialogue Box

Tips:

In the saved files, what stores is only the storage path of edited information and footage file, not the file itself. Therefore, if the file is stored on another computer, it is necessary to store the footage file together. When opening the files on a new computer, the software will ask for footage links, then re-link footage before you can edit the files.

Chapter 5 Application of Materials on PowerPoint of Multimedia Platform

The multimedia material processing softwares mainly realize the processing and production of a certain kind of multimedia material. In order to make multimedia works by comprehensively using a variety of multimedia materials around a specific theme, it is also necessary to make logical combination of various materials on the multimedia platform software to complete the final multimedia works.

Section 1 Introduction to Multimedia Platform Software

In general, when making multimedia works, we first use special multimedia material processing software to process multimedia materials such as image, animation, audio, and video, and then use multimedia platform software to organize these materials to form a whole one interrelated with each other. In addition, the multimedia platform software can also provide the production of operation interface, user interaction control, data management, and other functions.

Ⅰ. Types of Multimedia Platform Software

There are many kinds of software that can provide multimedia platform functions, including professional softwares applied for programming languages with high-level and connecting multimedia materials, and comprehensive software that can not only calculate but also process multimedia materials. The common multimedia platform softwares include Visual Basic, Authorware, PowerPoint, and so on.

1. Visual Basic

Visual Basic is a general object-oriented programming language developed by Microsoft Corporation. It originates from basic programming language and is usually referred to as VB for short. It has a graphical user interface and a program developing system with rapid application. The language completes the connection and calling of multimedia materials and the making of interactive programs through multimedia controls. The main workload of developing multimedia products with this language is programming. Developing program with multimedia products is effective and flexible. However, it puts forward high requirements for multimedia developers.

2. Authorware

Authorware is an interpretive, flow based graphic programming language, which can integrate sound, text, graphics, simple animation, digital video and other multimedia materials. The software is easy to operate, interactive and powerful. It has a large number of system functions and variables, so it is easy to realize program jump and reorientation. The whole development process of multimedia program can be carried out on the visual platform of the software. The structure of the program module is clear and concise. It can easily organize and manage each module by dragging the mouse, and design the calling relationship and logical structure between the modules.

3. PowerPoint

PowerPoint is one of a series of office softwares developed by Microsoft Corporation. It is mainly used to create presentation. Generally, the multimedia presentation product created by this software is called PPT. The design and production of PPT multimedia works can be realized with basic computer knowledge rather than professional programming ideas and means. Therefore, it is relatively easy at the very beginning. However, it is not easy to make an excellent PPT work. It needs to be familiar with and operate the software fluently, and design and integrate the content that needs to be displayed.

Ⅱ. Functions of Multimedia Platform Software

Multimedia platform software is an important supporting tool for the production of multimedia works. Its main functions are as follows:

①Control the start, run and stop of various media.

②Coordinate the time sequence between media, and carry out timing control and synchronization control.

③Generate the user oriented operation interface, and set the function menu of control button box so as to control the media.

④Generate and manage database.

⑤Monitor the operation of multimedia programs, including counting, timing, counting the number of events.

⑥Accurately control the input and output modes.

⑦Package the multimedia object program, set the installation file and unload file, and monitor and manage the environment resources and multimedia system resources.

It should be noted that it is unnecessary for each multimedia work to be possessed with the above functions. A specific multimedia work may only be applicable to a part of the above functions of the multimedia platform software.

Section 2　Insertion and Editing of Materials

PowerPoint is a presentation software released by Microsoft Corporation, which is a member of the Office family. PowerPoint can not only create presentations, but also hold face-to-face meetings and remote meetings on the Internet, which is widely applied in teaching, business reporting and other fields. Multimedia materials play an increasingly important role in PowerPoint production, which can make PPT content more visualized and specific, and marked with the imprint of times. The media materials in PowerPoint mainly include images, audio, video, etc. The following chapter takes PowerPoint 2016 as an example to introduce the application of multimedia materials.

I. Image Material

1. Insert Image

(1) Insert Background Picture

The background of a slide is usually determined by the template you select when you create the slide. The background of all the slides are blank by default. Sometimes the user needs to insert the background picture to change the background of a slide to achieve a distinctive effect. The steps are as follows:

①Open the slide to which you want to set the background picture, right-click in the blank space, select 【Format Background】, or click 【Set Background Format】 button in the Customize function area under "Design", as shown in Figure 5-1. The Format Background dialog box will pop up, as shown in Figure 5-2.

Figure 5-1　Format Background Button

Figure 5-2　Format Background dialog box

②In the Format Background dialog box, click Picture or texture fill, as shown in Figure 5-3, and click File to pop up the Insert Picture dialog box.

③Select the background picture to be inserted, as shown in Figure 5-4, and click the 【Insert】 button, and the effect is shown in Figure 5-5.

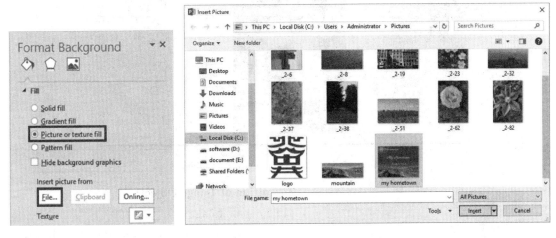

Figure 5-3 Picture or texture fill radio button and File button

Figure 5-4 Select Picture

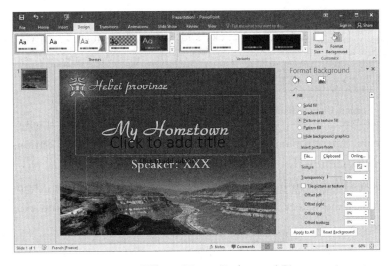

Figure 5-5 Effect of Insert Background Picture

(2) Insert Image Object

①Sometimes it is necessary to add the logo of the school or the company to the slide. At this time, it is necessary to insert the image from the file in the slide. The specific operation steps are as follows:

②Click the 【Picture】 button in the Image function area under 【Insert】 menu to pop up the Insert Picture dialog box, as shown in Figure 5-3.

③Select the image to be inserted, as shown in Figure 5-4. Click the "Insert" button, and the effect is shown in Figure 5-6.

Figure 5-6 Effect of Insert Image Object

Figure 5-7 Format Picture Panel

2. Editing of Image Object

(1) Format Image

After inserting the image into the slide, in order to improve the appeal of the picture, we can change the picture by setting the style of the picture, so that the picture looks more beautiful and attractive, and the effect can be highlighted. Select the picture, right-click, and select 【Format Picture】 in the pop-up menu, then the function panel will pop up, as shown in Figure 5-7.

1) Fill and Line

Click the 【Fill and Line】 function button in the Format Picture function panel to pop up the Fill and Line setting options. Users can set the line color of the picture through the Line option.

Click the 【Line】 function button, select Solid Line as the line type, Orange as the color, and 5 as the width, and select From Thick to Thin for the composite type. The effect is shown in Figure 5-8.

2) Effect

Click the 【Effect】 function button in the Format Picture function panel to pop up the Shadow, Image, Light and other effect options. The visual effect of the picture can be adjusted by setting the parameter value of each effect option. For example, click the "shadow" function button to set the shadow parameters as follows: select "Perspective Diagonal Upper Right" in the drop—down panel of "Presets" options, select the gray color "Gray—25％, Background —2, dark—75％" in the drop—down panel of Color option to set the color of the shadow, and set the Transparency option to 80％. The effect is shown in the Figure 5-9.

Chapter 5   The Application of Materials on PowerPoint of Multimedia Platform   · 115 ·

Figure 5-8   Effect of Picture Line Setting

Figure 5-9   Effect of Picture Shadow Setting

3) Size and Properties

Click the 【Size and Attribute】 function button in the Format Picture function panel to pop up the Size, Position, Text Box, Optional Text and other effect options. You can adjust the size and position of the picture by setting the parameter value of each effect option. Take setting the image size as an example, click the 【Size】 button, set the rotation to 30°, and set the zoom height to 35%. Select the Lock aspect ratio check box, and the effect is shown in Figure 5-10.

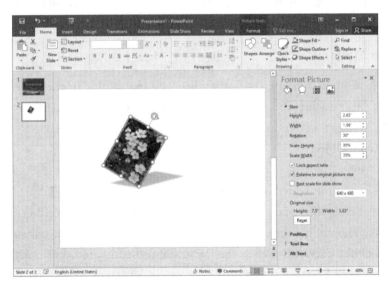

Figure 5-10   Effect of Picture Size and Properties Setting

4) Pictures

Click the 【Picture】 function button in the Format Picture function panel to pop up the Picture Correction, Picture Color, Clipping and other effect options. Users can adjust the color of the picture and crop the picture by setting the parameter value of each effect option. Take setting the image color as an example, click the 【Picture Color】 button, and preset the saturation as 300%. The effect is shown in Figure 5-11.

Figure 5-11   Effect of Picture Color Setting

(2) Set Transparent Color

In PowerPoint 2016, users can remove the background color of the picture through the Set Transparent Color function. Click the picture, click the 【Color】 function button in the 【Format】 menu, click 【Set Transparent Color】 in the pop-up menu, move the mouse to the background color of the picture, click the left mouse button, and the corresponding background color will become transparent.

II. Audio Material

In the production of slides, users can add various sound files to the slides to make them more appealing. After adding audio, a sound icon appears on the slide. The following chapter introduces how to insert a sound object into a slide.

1. Insert Audio Object

(1) Insert Audio File

Select the slide to be added with a sound file, click the 【Audio】 button in the media group of the insert ribbon, as shown in Figure 5-12. Select the 【Audio on My PC …】 from the drop-down list, as shown in Figure 5-13. The Insert Audio dialog box pops up, where users set the location of the audio file, select the audio file to be inserted, and finally click 【Insert】 button to insert a sound button into the slide.

After inserting a sound file, users will see a sound icon on the slide in the editing panel. Click the icon, and a simple playback controller will appear below the icon. The Audio tools area will appear in the tab area of the current window. Click the Format and Play tabs to make detailed settings for the inserted sound file.

Figure 5-12 Audio Button

Figure 5-13 Drop-down List of Audio Button

(2) Add Recorded Audio

Select the slide to be added with a sound file, click the 【Audio】 button in the Media group of the Insert ribbon, as shown in Figure 5-12. Select 【Record Audio】 from the drop-down list to open the Record Sound dialog box, as shown in Figure 5-14. Edit the file name and click the circle button to start recording. After recording, click the button to stop recording. Finally, click 【OK】 to insert the newly recorded audio file into the slide. The editing of this file is consistent with the previous audio file.

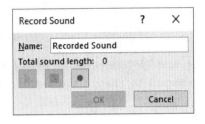

Figure 5-14 Record Sound Dialog Box

2. Editing of Audio Object

After inserting a sound file, users will see a sound icon on the slide in the editing panel. Click the icon to display a simple playback controller below the icon, as shown in Figure 5-15. In addition, the Audio Tools area appears in the tab area of the current window. Click the Format tab and the Play tab to make detail settings of the inserted sound file.

Figure 5-15 Playback Controller

(1) Audio Clip

After inserting audio into the slide, we can cut and play any file in the audio file without playing the entire audio file. The operation method is as follows:

Figure 5-16 Trim Audio Dialog Box

① Select the 【Audio Tools】 button in the function tab area, and click the 【Trim Audio】 button in the Editing group in the Playback ribbon to pop up the Trim Audio dialog box, as shown in Figure 5-16. There are two cutting display marks on the top of the audio file. The green one is to adjust the start position of the playback, and the red one is the end position. After setting the start and end time of the audio, click the 【OK】 button.

(2) Audio Playback Control

Select the inserted audio file and click the 【Audio Tools】 button in the function tab area. Audio playback can be controlled through the function options in the Audio Options group in the Playback function area, as shown in Figure 5-17.

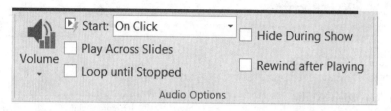

Figure 5-17   Audio Options Group

The default playback mode of the inserted audio is On Click. Click the Start drop-down menu option to change the video playback mode to Automatically playback, as shown in Figure 5-18.

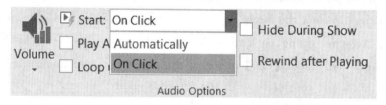

Figure 5-18   Set Playback Mode of Audio

Click the volume button in the Audio Options group to pop up the Volume options menu. Set the volume of audio playback as required. Set the sound volume to High, as shown in Figure 5-19.

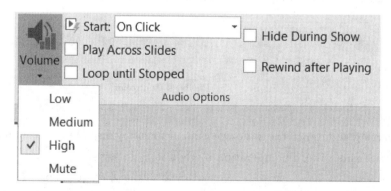

Figure 5-19   Set Audio Volume

In the Audio Options group, there are four check boxes that function as follows:

① Play Across Slides: if this check box is selected, audio can be played across slides;

② Hide During Show: if this check box is selected, the audio file icon will be hidden during audio playback;

③ Loop until Stopped: if the check box is selected, the audio will continue to play in a loop until the stop button is pressed;

④ Rewind after Playing: if this check box is selected, the audio will return to the beginning of the audio after playing.

Ⅲ. Video Material

In PowerPoint 2016, users can not only add images and sound effects, but also add videos, making slides more vivid and interesting. Inserting a video into a slide is similar to inserting a sound.

1. Insert Video Material

Select the slide to insert the video, click the 【Video】 button in the Media group of the Insert ribbon, select 【Video on My PC】 from the drop-down list, and the Insert Video dialog box will pop up, as shown in Figure 5-20. In the pop-up Insert Video dialog box, set the address to store the video file, select the video to be inserted, and click the 【Insert】 button to add a video to the slide.

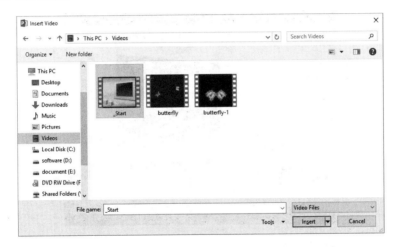

Figure 5-20   Insert Video Dialog Box

2. Editing of Video Object

(1) Video Clip

After inserting the video into the slide, we can cut the file at any position in the video file to play without playing the whole video file. The operation method is as follows:

①Select the 【Video Tools】 button in the function tab area, and click the 【Trim Video】 button in the edit group of the Play ribbon, as shown in Figure 5-21. The Trim Video dialog box will pop up, as shown in Figure 5-22.

②In the pop-up Trim Video dialog box, there are two clipping display marks at the bottom of the video preview panel. The green one is to adjust the starting position of the playback, and the red one is the ending position. Drag the mark to set the start time and end time of the video, and click the 【OK】 button to complete the video clipping.

(2) Video Playback Control

Select the inserted video file, and then click Video Tools and Playback. In the Video Options function area, the video playback can be controlled, as shown in Figure 5-23.

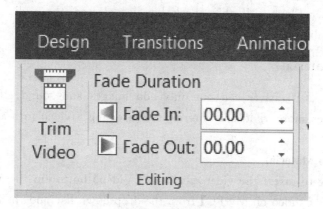

Figure 5-21 Trim Video Button

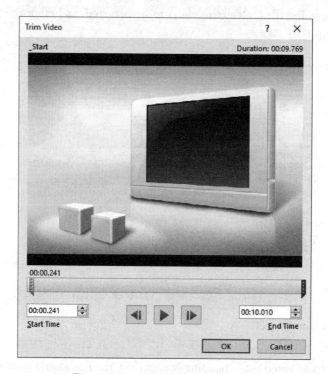

Figure 5-22 Trim Video Dialog Box

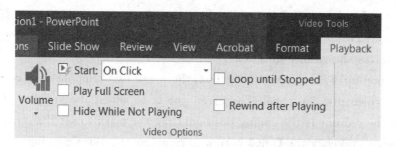

Figure 5-23 Video Options Group

The default playback mode of the inserted video is On Click. Click the Start drop-down menu option to change the video playback mode to Automatically playback, as shown in

Figure 5-24.

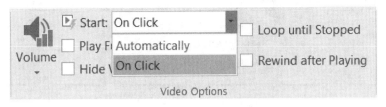

Figure 5-24 Set Playback Mode of Video

Click the volume button in the Video options group to pop up the Volume options menu. Set the volume of video playback as required. Set the sound volume to High, as shown in Figure 5-25.

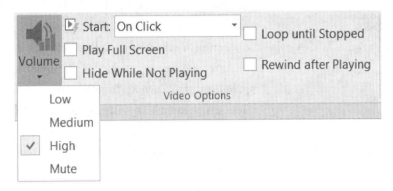

Figure 5-25 Set Video Volume

In the video options group, there are four check boxes that function as follows:

①Play Full Screen: if this check box is selected, the video will be played in full screen;

②Hide While Not Playing: if this check box is selected, the video will be hidden when it is not played;

③Loop until Stopped: if this check box is selected, the video will continue to play in a loop until the stop button is pressed;

④Rewind after Playing: if this check box is selected, the video will return to the beginning of the video after playing.

Ⅳ. Comprehensive Applications of Materials

When there are more than one picture in the slide, align the pictures and add animation effect, so that there is a hidden line between the pictures to connect them, and the page content is orderly placed, which will greatly enhance the expression of the slide.

1. Material Placement

(1) Overlaying Order

When multiple objects overlap, you can adjust the overlaying order of each image. To move the picture up one layer as an example, the operation method: select the image that needs to adjust the overlaying order, and click the 【Bring Forward】 button in the Arrange group of the Format function area of the image tool, as shown in Figure 5-26. You can change the overlaying order of pictures.

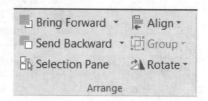

Figure 5-26　Function Button of Arrange Group

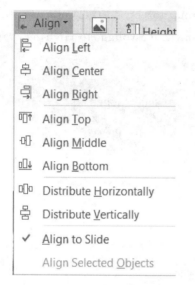

Figure 5-27　Drop-down List of Align

(2) Arrangement and Distribution

When inserting multiple picture objects, it can be very messy. The Align function of the picture tool can arrange all the pictures Vertically Centered and divided into two groups of Horizontal Distribution. The operation method is as follows: select 2 or more image objects and click the 【Align】 button in the Arrange group of the Format function area of the image tool. The drop-down list will pop up, as shown in Figure 5-27. Select an alignment method to complete the arrangement and distribution of multiple objects.

(3) Group

In PowerPoint 2016, you can use the Group function to combine several image objects to perform relative position scaling, moving and other operations as a whole. The group operation method is as follows: select 2 or more image objects and click the 【Group】 button in the Arrange group in the Format function area of the image tool. Select Group in the pop-up drop-down list, as shown in Figure 5-28. The combination of multiple objects can be completed. To cancel, select Ungroup from the drop-down list, as shown in Figure 5-29.

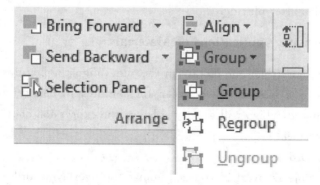

Figure 5-28　Function Button of Group

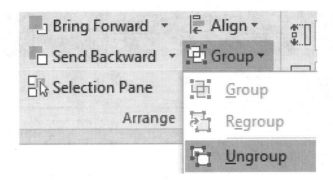

Figure 5-29　Function Button of Ungroup

2. Timing Control of Material Playing

The pictures, videos and other materials inserted into the slide can be controlled by adding animation effects. If the animation effect is set for each object on the slide, the objects in the slide are not displayed all at once, but are displayed in the way of animation in the order of setting.

(1) Types of Object Animation

There are four kinds of animation effects for the objects on the slide: entry effect, emphasis effect, path motion effect and exit effect.

①Entry Effect

The entry effect of an object refers to the animation effect when the object enters the projection interface during the slide show.

②Emphasis Effect

After the animation effect of the objects in the slide is set, the emphasis effect can be set to highlight some contents and increase the expressive force.

③Action Path Motion Effect

Action Path motion effect means that the objects in the slide can enhance the presentation effect of the slide according to the animation effect of the user specified path.

④Exit effect

Exit effect refers to the animation effect when an object exits the screen during a slide show.

(2) Object Animation Settings

①Set object entry effect

The operation steps for setting the entry effect of an object are as follows:

• Select the image object to set the animation effect, and click other buttons in the Animations group of the Animation functional area, as shown in Figure 5-30. You can also click the 【Add Animation】 button .

• The pop-up panel displays all preset animation schemes of PowerPoint 2016, as shown in Figure 5-31. Select the desired animation scheme in the Enter area to complete the animation of the object. If you are not satisfied, you can click the "More Entrance Effects" option at the bottom of the panel to make more detailed settings in the pop-up Add Entrance

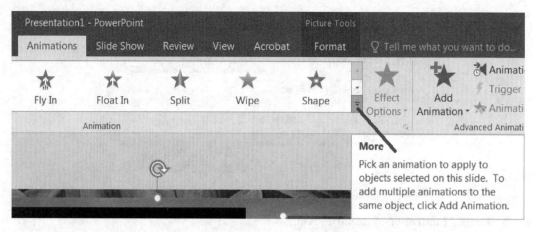

Figure 5-30 Other Buttons of Animations Group

Effect dialog box, as shown in Figure 5-32.

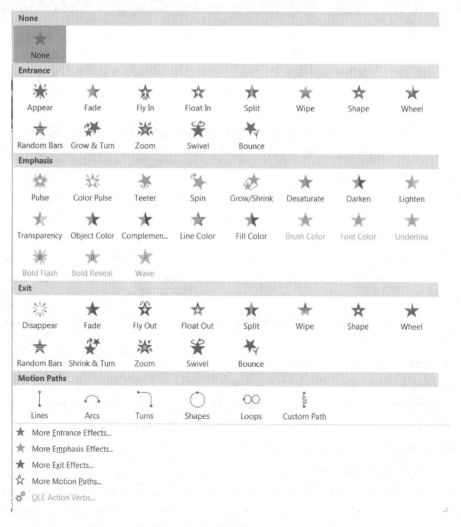

Figure 5-31 PowerPoint2016 Preset Animation Schemes

- In the Animations group, click the 【Effect Options】 button and select the appropriate direction button in the drop-down list to set animation for the selected object, as shown in Figure 5-33.

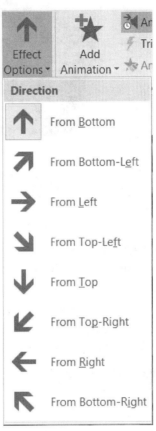

Figure 5-32　Add Entrabce Effect Dialogue Box　　　Figure 5-33　Effect Options Setting

②Set Object Emphasis

Select the object that needs to set the Emphasis effect. In the animation scheme panel, as shown in Figure 5-31, select an animation scheme in the Emphasis area to complete the object emphasis effect setting.

After the object has been animated with emphasis, click the 【Effect Options】 button in the Animations group to set more detail effects.

③Set the motion effect of objects according to Action Path.

Select the object that needs to set the motion effect of Motion Paths. In the animation scheme panel, as shown in Figure 5-31, select an animation scheme in the Motion Paths area, and click 【OK】 to set the motion paths effect for the selected object.

The set Motion Paths will appear on the slide. It supports rotation, reshape and resize using the control block as shown in Figure 5-34, so as to diversify the action path of the object.

④Sets the Exit Effect of an object

Select the object to set the Exit Effect. In the animation scheme panel, as shown in

Figure 5-34　Motion Path Setting

Figure 5-31, select an animation scheme in the Exit area, and click 【OK】 to set the exit effect for the selected object.

(3) Add, set, and delete animations

①Add animation

Users can set multiple animation effects for each object on the slide. The operation method is as follows: select the object that has been animated, and click 【Add Animation】 button in the Advanced Animation group of the Animation ribbon to add animation effect to the object again.

②Set Animation

As long as any object on the slide is animated, you can select the object again to further set its animation effect. Operation method: select the object and click the button in the timing group of the Animation ribbon to set the animation start time, duration and delay time.

When you want to change the order of object animation, you can click the animation pane button in the Advanced Animation group, and the animation pane appears on the right side of the slide editing pane, as shown in Figure 5-35. The animation set on the current slide is displayed in the animation pane. After selecting the animation line marked by numbers in the animation pane, a drop-down button appears on the right side of the line. Click the drop-down button to realize the detailed setting of the animation, including animation start conditions, effect options, timing, deletion, etc. If you select an animation in the animation pane, the upper and lower arrow on the right side of the pane is in the available state, which is used to adjust the order of animation. If you click the up arrow button, you can move the selected animation sequence one level.

③Delete Animation

There are two ways to remove animation effects from slides:

- Select the object, click the animation to be deleted, and press the 【Delete】 key;
- Select the object, click the drop-down button on the right side of the Animation Line to be deleted in the animation pane, and select the Remove option in the drop-down option, as

shown in Figure 5-36, to delete the animation.

Figure 5-35   Animation Pane

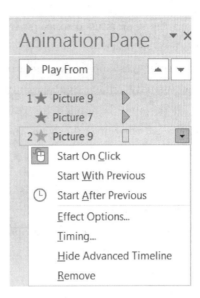

Figure 5-36   Delete Animation

Section 3   Slide Design

The quality of design is relied on many factors, including inspiration, conception, layout, structure, visual presentation and so on. The slide design introduced in this chapter is only the tip of the iceberg, but its design principles and methods are similar to other types of design.

I. Frame Design

Slide frame design is the first step of slide design. It mainly designs the overall structure of slide, including the overall logic of slide and the content page type of slide.

1. Logic design

People's understanding and perspectives on the same content might be different because of the difference of the speakers' illustration ways rather than that of the audiences' understanding capabilities, which lead to the diversities of easy to understand and difficult to understand. Logic means what you want to say first and what to say next; which is important and which is secondary; what you want to remember and what you ignore.

Slide conception and design depend on the purpose of the slide. The conception way of slides for auxiliary presentation is different from that for reading, and the design way of slides are diversified due to operating experiences of various users. Therefore, it is necessary to consider the use of slides and audience types, and conceive from the whole to design satisfactory slides.

> Tips:
>
> If the slide is used to assist the presentation, and there are speakers on the stage, the slide can be based on pictures, with more real high-definition pictures to stimulate the audience's eyes, and at the same time, it can arouse the audience's emotional resonance, which usually achieves good results.
>
> If the slide is applied for reading, it is necessary to ensure enough text to provide enough information; otherwise, the audience will not get the desired effect with a little understanding.

In fact, the logic of slides is the logic of speech, which includes different ways and methods such as sequence, causality, importance, deduction and induction. Constructing logic is to present the content to the audience in a well-designed structure and make it easy for them to understand and accept.

2. Page Planning

After confirming the logical structure of the slide presentation, the next step is to plan the page structure. A complete presentation generally includes the following page types:

Cover page: the first slide in a set of slides, usually containing the title of the speech and speaker's information.

Abstract page: give a general description of the whole speech, which is a bit like the summary in a book. The abstract page can be selected according to the speaker's arrangement.

Table of contents: similar to the contents of books, it lists the outline of speech content in the form of title, so as to facilitate the audience to master the overall structure of the speech. Users can also leave the table of contents out of the slide.

Transition page: When the content of a speech is switched from one aspect to another, there is a clear transition page to let the audience know clearly that the previous part has been finished. Now it is convenient for the audience to follow the ideas of the presenter.

Content page: including the main content of the slide, is the core part of the slide.

Summary page: after the explanation of all the contents, a general summary of all the previous contents can be made. According to the needs of the speech and the length of time, the summary content can be more or less.

Closing page: the last page of the whole slide, usually leaving the contact information of the speaker or the words of thanks on the closing page.

Among the above page types, the cover and content page are indispensable. The transition page can prompt the progress of the speech. Generally, it should be added. The closing page also reminds the audience of an obvious ending, which is generally needed.

II. Page Design

The effect of page design depends on the design style and speech content of slide designer. The effects of slide design with different designers and different themes are also different. However, the design of slides should generally grasp the following points:

1. What matters is not the quantity but the quality of the content.

Since the space of a PPT is limited, don't write everything but the key points on the page. Not only the text and picture area, but also appropriate blank space are very necessary, so that people's eyes will not be tired. A slide with high quality should be carefully selected, and appropriate in content, which can reflect the speaker's central idea or point of view.

2. What matters is not the quantity but the harmonious matching of colors.

Beginners often make two common mistakes: one is the misuse of color, which gives people a sense of disorderly pages; the other one is the non-use of color, that is, the only application of black fonts from the beginning to the ending page. It's allowed to adopt different colors, but they should be harmoniously arranged. How to achieve harmony? It is to determine the main color. The general rule is: for the bottom plate with light color, the main color is light color, and the text is dark color; for dark bottom plate, the main color is transparent or light color, and the text is light color. According to experience, light color on the bottom plate is easier to combine and match colors than dark one. After the main tone is determined, add a darker color text, so that the text can be more prominent. Avoid the reversal of the order of main color and sub-color.

3. What matters is not the quantity but thenecessity of animation.

The addition of animation effect is the difference between film slide and multimedia slide. Appropriate and exquisite animation is undoubtedly an eye-catching tool. But it can be imagined that inappropriate or excessive animation will be equally offensive. One thing to remember is that not every animation effect in the list is suitable for your slide. Usually, less than 10 animations are applied on the slide, but skillfully using the combination of animations will generate infinite effects.

4. Three requirements: less words, fewer formulas and larger fonts

These three points are the basis of slide page design, which are not only related to the beauty of the slide, but also related to the usability of the slide. The auxiliary presentation slide can't display all the speech words on the slide page, because people's visual sensitivity is much higher than the hearing, and all the contents of the speech are on the slide, the audience will naturally pay more attention to the text reading, but ignore the speaker's explanation. Therefore, when designing presentation slides, it is not suggested to collect all the speech words on the side as an assisting method. Slides for reading can be written with a little more text, but also can't full screen text, which will bring weariness to readers.

Ⅲ. Design Tips for Beginners

As a beginner, how to make a good slide works when the slide design is not easy to be managed? The answer is "on the shoulders of giants". PowerPoint software provides us with a very convenient slide design mode, which is divided into three sections: design template, color scheme and switching scheme.

1. Design Template

The design template is the template you want to apply in this slide. Among them, the slide background is the most concerned. PowerPoint has a lot of templates, including many practical templates. In the daily production of slides, these templates have been able to cope with. To convert the applied template, you just need to click the corresponding pattern in the design template. It is worth noting that the general template has its own unique format, such as font and color, so it is necessary to carefully check whether the slide is displayed normally after the template transformation.

2. Color Scheme

Color scheme refers to the color combination of foreground color, background color, emphasis content color and hyperlink color used in slide. The general PowerPoint template will provide automatically calculated color scheme for selection. The most important one is fill color, which determines the default fill color of all your custom graphics. Slide makers can also modify the color scheme, but for beginners or producers with weak art foundation, it is recommended not to modify the color scheme.

3. Animation Scheme

Animation scheme is also known as slide transition, that is, the transition animation in each slide show interval. Beginners often ignore this function. In fact, this is the easiest way to make slides colorful. As long as you choose the appropriate scheme, your PPT will show advanced animation effects such as fade in and fade out when switching, making your slides more professional. You can choose to apply an animation effect to a single slide or a whole slide, so that your display will be rich and colorful, especially suitable for slides of similar types of people, products or activities. Although the animation effect looks very dazzling, but if you add too much, or the animation selection is not appropriate, it will be counterproductive. For example, if the overall use of "cube", "library" and other switching animation with strong picture impact, the audience will feel dizzy; if there are pictures of characters on the slide, it is better not to use such switching effects as "random lines" and "blinds" that give people a sense of splitting.

Two principles need to be grasped in the animation scheme design of the slide: one is that the animation scheme should have a logical leading role. For example, a unified animation with strong visual impact, such as "Cube", "library" and "door", should be added to each transition page to attract the attention of the audience, so that the audience can keep up with the conversion of the speaker's ideas; the other is that the animation scheme should be appropriate, for example, it is not recommended to make animation with split effect.

For beginners, in addition to using PowerPoint's own design scheme, you can also find many excellent slide design cases through the Internet, such as VIP, Microsoft official website, etc., from which you can learn the design methods and skills of excellent designers, and also learn from the design inspiration to complete their own slide design.

References

[1] Shi Weiming. Research Studio 16 Lessons of Mastering PHOTOSHOP[M]. Beijing: China Machine Press2010.07.

[2] Jin Hao. From being a green hand to a professional PHOTOSHOP Chinese edition Beijing:China Machine Press2012.05.

[3] Liu Haiying. Digital Audio Processing Tutorial[M]. Beijing:Tsinghua University Press 2020.11

[4] Anonymity GoldWave Digital Audio Editing Tutorial[EB/OL](2021-08-06)[2021-12-05]. http://wk.baidu.com/view/e1162c77e73a580216fc700abb68a98270feac53.

[5] Lu Shan Culture. Corel Video Studio X9 Video Editing and Producing[M]. Beijing: China Machine Press. 2017.11

[6] Zhao Zijiang. Multimedia technology application tutorial[M]. Beijing:China Machine Press. 2010.02